厦门大学海洋法与中国东南海疆研究中心
厦门大学南海研究院海洋事务系列丛书

傅崐成 ◎ 主编

现代科技背景下
海洋法中的海洋科学研究

Marine Scientific Research
under the International Law of the Sea in the Era of Marine High-Tech

常 虹 ● 著
By Hong Chang

厦门大学出版社 国家一级出版社
XIAMEN UNIVERSITY PRESS 全国百佳图书出版单位

The present study has been accepted as PhD thesis by the Department of Law of Trier University, Germany in the summer term 2013. The two assessors are Professor Alexander Proelss and Professor Gerhard Robbers.

Acknowledgements

First and foremost I would like to thank my supervisor, Alexander Proelss. This thesis profited immensely from his helpful guidance, comments and suggestions. As a foreign student, my English is not always so satisfactory and the way of legal reasoning is also sometimes different. Professor Proelss is always patient and gives me very constructive comments. I am especially impressed by his sharp perception and far-sighted ideas. I therefore express my deep gratitude to him.

I would like to thank Professor Andreas Oschlies, Professor Arne Koertzinger and Tobias Steinhoff from GEOMAR. I very much appreciate their guidance in helping me with understanding the basic marine scientific knowledge, introducing problems and gathering ideas. I would like to thank Professor Martin Visbeck from GEOMAR for his support in many ways.

Furthermore, my sincere gratitude goes to Petra Gnadt, secretary of Professor Proelss and now secretary of Professor Nele Matz-Lueck in Walther-Schücking-Institut für Internationales Recht. I am so glad that I could meet her four years ago. In these four years, she has helped me in many ways with all the best heart.

I wish also to acknowledge Sylvia Nwamaraihe from Walther-

Schücking-Institut für Internationales Recht and Aaron Lemkow from Dalhousie University in Canada for language checking and polishing and their helpful comments.

The friendship with Kate Houghton accompanied me throughout my four years' study. I therefore gratefully thank her for all her help and smile.

I would like to thank China Scholarship Council for granting me four years' scholarship that enables me to complete my studies. My thanks also go to Walther-Schücking-Institut für Internationales Recht. It provides me a very supportive and inspiring research environment with nice colleagues and comprehensive access to literatures. I am grateful to the Kiel Cluster of Excellence "The Future Ocean". Not only it provides financial support for my study in Canada and some conferences I attended, but also offers a broad interdisciplinary network which helped me to understand the basic scientific knowledge.

The final words of appreciation are reserved for my family. My parents deserve the deepest thanks for their support, patience and trust. My grandma, I will miss her forever.

Contents

I. Introduction 1
   A. Marine Scientific Research in the Era of Marine High-Tech 1
   B. The International Law of the Sea—A Brief Overview 8
   C. Problem Identification 17
   D. Research Questions 20
   E. Structure of the Book 22

II. Marine Scientific Research under the International
    Law of the Sea 24
   A. Prior to 1958 24
   B. 1958 the First UN Conference on the Law
      of the Sea(UNCLOS I) 29
      1. MSR in the High Seas 30
      2. MSR on the Continental Shelf 32
   C. The 1982 United Nations Convention on the Law of the Sea 41
      1. Sovereignty—Territorial Sea 42
      2. Jurisdiction—MSR in the EEZ and on the Continental Shelf 49
      3. Freedom—MSR in the Areas beyond National Jurisdiction 56
      4. Hydrographic Surveys 61
      5. 1982 LOS Convention Article 246 66
   D. Conclusion 90

III. Implementing the MSR Regime in the Marine High-Tech Era 93
   A. Ocean Upwelling 94
      1. Introduction 94
      2. Legal Analysis 97
      3. Conclusion 118

B. Voluntary Observing Ships ... 119
1. Factual Background ... 119
2. Legal Analysis ... 125
3. Conclusion ... 155
C. Marine Bioprospecting ... 157
1. Factual Background ... 157
2. What Is Meant by "Bioprospecting" ... 159
3. Legal Assessment ... 164
4. Conclusion ... 202
Ⅳ. Final Remarks ... 205
Bibliography ... 208

List of Abbreviations

ABE-LOS	Advisory Body of Experts on the Law of the Sea
ABNJ	Area Beyond National Jurisdiction
AUV	Autonomous Underwater Vehicles
CBD	Convention on Biological Diversity
CCS	(Geneva) Convention on the Continental Shelf
CDR	Carbon Dioxide Removal
CTS	(Geneva) Convention on the Territorial Sea and the Contiguous Zone
CHS	(Geneva) Convention on the High Seas
EEZ	Exclusive Economic Zone
etc.	et cetera
et seq.	et sequentes
GTS	Geostationary Technology Satellite
ibid	idem (the same)
ICNT	Informal Composite Negotiating Text
ICSU	International Council for Science
IHO	International Hydrographic Organization
ILC	International Law Commission
IOC	Intergovernmental Oceanographic Commission
ISA	International Seabed Authority
ISNT	Informal Single Negotiating Text
ITLOS	International Tribunal on the Law of the Sea
IUU fishing	Illegal, Unreported and Unregulated fishing
JCOMM	Joint IOC-WMO Technical Commission on Oceanography and Marine Meteorology
London Convention	Convention on the Prevention of Marine Pollution by Dumping of Wastes and Other Matter 1972
London Protocol	Protocol to the London Convention

LOS Convention	(United Nations) 1982 Law of the Sea Convention
MARPOL	International Convention for the Prevention of Pollution from Ships
MSP	Marine Spatial Planning
MSR	Marine Scientific Research
NMHS	National Meteorological and Hydrological Service
NOAA	National Oceanic and Atmospheric Agency
ODAS	Ocean Data Acquisition Systems
OECD	Organization of Economic Cooperation and Development
OF	Ocean Fertilization
OPA	Observations Programme Area
pCO_2	Carbon Dioxide Partial Pressure
PMO	Port Meteorological Officer
SBSTTA	Subsidiary Body on Scientific, Technical and Technological Advice
SOT	Ship Observation Team
SRM	Solar Radiation Management
TSG	Thermosalinograph
UK	United Kingdom of Great Britain and Northern Ireland
UN	United Nations
UNCLOS Ⅰ	First United Nations Conference on the Law of the Sea
UNCLOS Ⅱ	Second United Nations Conference on the Law of the Sea
UNCLOS Ⅲ	Third United Nations Conference on the Law of the Sea
UNESCO	United Nations Educational, Scientific and Cultural Organization
UNFCCC	United Nations Framework Convention on Climate Change

◇List of Abbreviations◇

USSR	Union of Soviet Socialist Republics
VOS	Voluntary Observing Ship
WMO	World Meteorological Organization
Working Group	Ad Hoc Open-ended Informal Working Group

I. Introduction

A. Marine Scientific Research in the Era of Marine High-Tech

Many people may have noticed that recently there was a controversial worldwide ocean iron fertilization project that was launched off the Canadian west coast in July 2012.[1] The project leaders claimed that they had violated neither Canadian regulations nor international law,[2] while lawyers, environmentalists and civil society groups spoke of "blatant violations" of international law[3] that could have serious repercussions. A few months later, a statement was made by the contracting parties to the London

---

[1] More information about the news available at http://www.nature.com/news/ocean-fertilization-project-off-canada-sparks-furore-1.11631 or http://www.guardian.co.uk/environment/2012/oct/15/ pacific-iron-fertilisation-geoengineering (last visited on 14 April 2013).

[2] http://www.cbc.ca/news/canada/british-columbia/story/2012/10/19/bc-ocean-fertilization-haida.html(last visited on 14 April 2013).

[3] http://www.guardian.co.uk/environment/2012/oct/15/pacific-iron-fertilisation-geoengineering(last visited on 14 April 2013).

Convention (LC)[④] and London Protocol (LP)[⑤], at their joint meetings in London from October 29th to November 2nd, 2012, which addressed the international concern of the iron fertilization project in ocean waters west of Canada.[⑥] It stated that:

1. The Parties to the London Convention and London Protocol (LC/LP) express grave concern regarding the deliberate ocean fertilization activity that was recently reported to have been carried out in July of 2012 in waters off the Canadian west coast.

2. [...]

3. The Parties to the London Convention and London Protocol reiterate, as agreed in 2008, that ocean fertilization activities, other than legitimate scientific research, should not be allowed.

4. The Parties reiterate that legitimate scientific research is defined as those proposals that have been assessed and found acceptable under the 2010 "Assessment Framework for Scientific Research Involving Ocean Fertilization".

5. [...]

By 2007, the contracting parties to the LC/LP had declared that ocean fertilization activities fell within the competence of both instruments, especially related to their obligation to protect the marine environment.[⑦] In

---

④ Convention on the Prevention of Marine Pollution by Dumping of Wastes and Other Matter 1972, London, 29 December 1972, in force 30 August 1975, 11 ILM (1972) 1294.

⑤ Protocol to the London Convention, London, 7 November 1996, in force 24 March 2006, 36 ILM (1997) 1.

⑥ The document sees http://www.imo.org/OurWork/Environment/SpecialProgrammesAndInitiatives/Pages/London-Convention-and-Protocol.aspx (last visited on 14 April 2013).

⑦ See Philomene Verlaan, Marine Scientific Research: Its Potential Contribution to Achieving Responsible High Seas Goverance, in: *The International Journal of Marine and Coastal Law*, Vol. 27, 2012, 805-812, at 807.

Ⅰ. Introduction

2008, ocean fertilization was defined[⑧] and essentially prohibited, though only in the form of a non-binding LC/LP Resolution.[⑨] There was an exception for an ocean fertilization-related category that was new to international law in general and the international law of the sea in particular: legitimate scientific research.[⑩] By virtue of the LC/LP Resolution 2008, "legitimate scientific research should be defined as those proposals [for ocean fertilization] that have been assessed and found acceptable under the assessment framework"[⑪]. The assessment framework, developed by the LC/LP's joint scientific groups and adopted in 2010,[⑫] provided a tool for assessing the proposed ocean fertilization activities on a case-by-case basis to determine if the proposed activity constitutes legitimate scientific research that is not contrary to the aims of the LC or LP.[⑬] Based on the assessment framework, two major assessments are required: an Initial Assessment and an Environmental Assessment.[⑭] The Initial Assessment, just as its name implies, determines whether a proposed activity is eligible to be considered and evaluated in this framework. Two qualifications are expressly brought forward: first, the proposed activity should fall within the definition of ocean fertilization and second, it should have proper scientific attributes.[⑮] In order to determine if a proposed activity has proper scientific attributes, it

---

⑧ It was defined as "ocean fertilization is any activity undertaken by humans with the principal intention of stimulating primary productivity in the oceans".

⑨ The document sees supra. note 6.

⑩ See Philomene Verlaan, supra. note 7.

⑪ See Resolution LC-LP. 1 (2008) on the Regulation of Ocean Fertilization, http://www.whoi.edu/fileserver.do? id=56339&pt=10&p=39373 (last visited on 15 July 2013).

⑫ Ibid. Resolution LC-LP. 2 (2010), to which the assessment framework is annexed.

⑬ See http://www.imo.org/blast/blastDataHelper.asp? data_id=30641&filename=AssessmentFramework-Annex6-LC-32-15.pdf (last visited on 14 April 2013).

⑭ Ibid.

⑮ Ibid.

should meet the following criteria:[16]

1. The proposed activity should be designed to answer questions that will add to the body of scientific knowledge. Proposals should state their rationale, research goals, scientific hypotheses and methods, scale, timings and locations with clear justification for why the expected outcomes cannot reasonably be achieved by other methods;

2. Economic interests should not influence the design, conduct and/or outcomes of the proposed activity. There should not be any financial and/or economic gain arising directly from the experiment or its outcomes. This should not preclude payment for services rendered in support of the experiment or future financial impacts of patented technology;

3. The proposed activity should be subject to scientific peer review at appropriate stages in the assessment process. The outcome of the scientific peer review should be taken into consideration by the Contracting Parties. The peer review methodology should be stated and the outcomes of the peer review of successful proposals should be made publicly available together with the details of the project. Where appropriate, it would be beneficial to involve expert scientists from other countries; and

4. The proponents of the proposed activity should make a commitment to publish the results in peer reviewed scientific publications and include a plan in the proposal to make the data and outcomes publicly available in a specified time-frame.

These criteria not only enunciate the notion of proper scientific attributes, but also "[constitute] the first detailed and sophisticated exposition, in an international instrument designed to guide states in the im-

---

[16] See http://www.imo.org/blast/blastDataHelper.asp? data_id=30641&filename=AssessmentFramework-Annex6-LC-32-15.pdf (last visited on 14 April 2013).

I. Introduction

plementation of their treaty obligations, and [it] also intended to become legally binding itself, of criteria that are also suitable for developing a general—i. e., not limited to OF [ocean fertilization]—definition of (marine) (scientific) research under the LOSC [United Nations Convention on the Law of the Sea]"[17].

The 1982 United Nations Convention on the Law of the Sea (LOS Convention),[18] known as the constitution for the oceans, stipulates in its Preamble that it is desirable to establish "through this Convention, ... a legal order for the seas and oceans which... will promote the... study... of the marine environment". That is said, the concept concerned then undergoes a name change, from "study of the marine environment", which appears only in the Preamble, to "research", usually—but not always—prefaced by two adjectives: "marine" and "scientific".[19] The LOS Convention devotes the entire Part XIII (Articles 238-265) to "marine scientific research". Among other LOS Convention articles, the term "research", with or without one or both of the aforementioned adjectives, features at least fifteen times.[20] Oddly enough, nowhere does the LOS Convention define "research" in any of its three incarnations.[21] This well-known absence of an internationally agreed definition for marine scientific research (MSR) has given rise to, in terms of ocean governance, a certain degree of difficulty that is occurring in a growing range of contexts. For instance, it is oftentimes questionable whether or not a MSR application needs to be made for collecting marine meteorological data and other routine ocean observations in light of the differentiation between "marine scientific

---

[17] See Philomene Verlaan, supra. note 7, at 808 (italics added).

[18] United Nations Convention on the Law of the Sea, Montego Bay, 10 December 1982, in force 16 November 1996, 1833 UNTS 397.

[19] See Philomene Verlaan, supra. note 7, at 806.

[20] Ibid.

[21] Ibid.

research" and "hydrographic surveys" represented in the LOS Convention Article 21.[22] Moreover, scientific ambitions to investigate the oceans, together with technical developments, have given rise to instruments and methods such as sea gliders and ice-tethered moorings that operate autonomously.[23] Such activities obviously differ from classic ship-based expeditions and are thus not easily addressed by the LOS Convention MSR regulations.[24] According to the US Department of State, the activities not amounting to MSR include:

> prospecting for and exploration of natural resources; hydrographic surveys (for enhancing the safety of navigation); military activities including military surveys; activities related to the laying and operation of submarine cables; environmental monitoring and assessment of marine pollution pursuant to section 4 of Part XII of the Convention; the collection of marine meteorological data and other routine ocean observations—such as those used for monitoring and forecasting of ocean state, natural hazard warnings and weather forecasts, and climate prediction—including through the voluntary ocean observation programs of the Joint Intergovernmental Oceanographic Commission—World Meteorological Organization Technical Commission on Oceanography and Marine Meteorology (JCOMM), the Global Drifter Programme, and the Argo programme; and activities directed at objects of an archeological and historical nature found at sea.[25]

Given the foregoing criteria that fit the concept of proper scientific attributes, it may offer some insights on how to define MSR. While it is not

---

[22] See Uwe Nixdorf, Arctic Research in Practice, in: Susanne Wasum-Rainer, Ingo Winkelmann and Katrin Tiroch (eds.), *Arctic Science, International Law and Climate Change*, Springer, 2011, 67-81, at 73.

[23] Ibid.

[24] Ibid.

[25] Marine Scientific Research Authorizations, available at: http://www.state.gov/e/oes/ocns/opa/rvc/index.htm (last visited on 5 August 2013).

◆ Ⅰ. Introduction ◆

simple to evaluate the consequences of the lack of a definition of MSR, it cannot be doubted that MSR is capable of affecting people's life in various ways. It is believed that research and applying the obtained expertise are the keys to modern ocean exploration and protection. Due to the fragile environment, the key issue is how to organize economic activities while taking environmental concerns into account and keeping the need for sustainable development as the fundamental perspective. Utilizing natural resources requires know-how, caution and a sense of responsibility. Besides, it should also be kept in mind that MSR outcomes have a key role to play in shaping ocean policy development. Policy-makers need to be informed of current scientific results, and the scientific findings should be conveyed on a regular basis to decision-makers in a manner that emphasizes their implications for making and implementing policies. Engaging marine scientists early during the policy-making process is thus essential to ensure that policy definitions and strategic directions are understood clearly and correctly. Historically, science and technology have been important triggers for legal development. For at least a half of the twentieth century, the law of the sea has looked to science and applied scientific tests in several different contexts, such as the conservation of fish stocks, the delimitation of the continental shelf, and the preservation and protection of the marine environment.[26]

It is believed that improving the knowledge and understanding of the ocean is a prerequisite for protecting the marine environment and ecosystems in a precautionary way and it promotes sustainable economic opportunities for ocean resources exploration or exploitation. Results from MSR will provide input for policy-makers in pursuing developmental options and also benefit society in terms of weather forecasting and preventing natural

---

[26] See David Anderson, Scientific Evidence in Cases under Part XV of the LOSC, in: Myron H. Nordquist, Ronan Long, Tomas H. Heidar and John Norton Moore (eds.), *Law, Science & Ocean Management* (Martinus Nijhoff Publishers, 2008), 505-518, at 506.

disasters. As much as science and technology plays an increasingly fundamental role in policy and law-making, it is most importantly a huge challenge to adapt to contemporary scientific knowledge and even anticipate the social change produced therein. [27] Regarding the relationship between the Law of the Sea and MSR, John Knauss once remarked that "it is simplistic to say that ocean policy, and therefore the Law of the Sea is driven by science and technology, but it is worth remembering"[28].

B. The International Law of the Sea—A Brief Overview

For the sake of better understanding the legal background regarding MSR within the context of the LOS Convention, it is necessary to bring forward a brief overview on the International Law of the Sea. D. P. O'Connell once expressed that: "[t]he history of the law of the sea has been dominated by a central and persistent theme: the competition between the exercise of governmental authority over the sea and the idea of the freedom of the seas."[29] Prior to the seventeenth century, the ocean was assumed as res nullius and therefore subject to unilateral appropriation. [30] Whenever one or two great powers dominated the international relations or achieved parity of power, emphasis was try to put as much ocean space as possible under

---

[27] See Aldo Chircop, Advances in Ocean Knowledge and Skill: Implications for the MSR Regime, in: Myron H. Nordquist, Ronan Long, Tomas H. Heidar and John Norton Moore (eds.), *Law, Science & Ocean Management* (Martinus Nijhoff Publishers, 2008), 575-615, at 580.

[28] Excerpted ibid., at 576.

[29] See D. P. O'Connell, *The International Law of the Sea*, Vol. 1 (Clarendon Press, 1982), at 1.

[30] See Bernard G. Heinzen, The Three-Mile Limit: Preserving the Freedom of the Seas, in: *Stanford Law Review*, Vol. 11, 1959, 597-664, at 598.

Ⅰ. Introduction

their control. This was the case when the Treaty of Tordesillas[31] was concluded between Spain and Portugal in 1494, which drew a line down a meridian of longitude through Brazil.[32] Pursuant to the Treaty, east of that line would be the area of Portuguese expansion and westward would be that of Spain.[33] Originally, this was not intended as a reservation of the seas to Spain and Portugal.[34] O'Connell expressed that: "it was later thought that it had that effect when both countries forbade trade within their respective areas. It was, in fact, to counter the supposed pretension of the Portuguese in the East Indies that Grotius wrote on the freedom of the seas early in the seventeenth century."[35]

In the second half of the sixteenth century, the King of Poland and the Queen Elizabeth Ⅰ of England initiated a new trend when declaring that "the use of the sea and air is common to all; neither can any title to the ocean belong to any people or private man, forasmuch as neither nature nor regard of the public use permitteth any possession thereof".[36] Accordingly, in the first half of the seventeenth century, a Dutch lawyer, Grotius, brought forward the "freedom of the seas" doctrine,[37] prescribing that the high seas areas communis or the common property of all nations.[38] The significance of bringing forward "common property of all nations" is that no State may unilaterally appropriate any portion of the high seas without the general consent of other States.[39] Against the historical context, it is

---

[31] Full image of the original kept by the Spanish Historical Archives at: http://www.mcu.es/archivos/docs/Documento_Tratado_Tordesillas.pdf (last visited on 22 July 2013).

[32] See D. P. O'Connell, supra. note 29, at 2.

[33] Ibid.

[34] Ibid.

[35] Ibid.

[36] See Bernard G. Heinzen, supra. note 30, at 599.

[37] See Donald R. Rothwell, Tim Stephens, *The International Law of the Sea* (Hart Publishing Ltd., 2010), at 3.

[38] See Bernard G. Heinzen, supra. note 30, at 600.

[39] Ibid.

believed that Grotius' "freedom of the seas" doctrine was originally part of a legal opinion on the question of Dutch East India Company's right of access to the trade of the Indies.⑩ Grotius argued that the freedom covered the whole of the oceans, and the seas were avenues of commerce which of their nature are not susceptible of appropriation.⑪

However, shortly after the turn of the seventeenth century, England temporarily reversed its advocacy of the freedom of the seas and claimed a mare clausum when Elizabeth's successor James Ⅰ suddenly claimed sovereignty over "our coasts and seas".⑫ It is therefore not surprising that the publication of Grotius' "freedom of the seas" doctrine in 1608 led to confrontation with Scottish and English jurists,⑬ especially from John Selden and William Welwood.⑭ It is believed that "[t]he controversy that captured the mind of scholars centers in the opposing doctrines of Grotius and Selden and it is until now the most celebrated and quoted as giving birth to the adversarial forces that shaped modern Law of the Sea into a balance between the freedom of the seas and the power of coastal states"⑮. Nevertheless, before long, with the trend of freedom of seas came back to prevail, England returned to the Elizabethan position.⑯

---

⑩ See D. P. O'Connell, supra. note 29, at 9.

⑪ Ibid.

⑫ See Bernard G. Heinzen, supra. note 30, at 600. James Ⅰ began to reverse the Elizabethan policy in 1604 when he proclaimed jurisdiction over the "King's Chambers" for purposes of neutrality. The chambers, twenty-six in number, were delineated by drawing straight baselines between headlands along the English coast. The largest chambers were on the west coastal where one baseline was nearly one hundred nautical miles long. Excerpted from Bernard G. Heinzen, supra. note 30, at note 11, 600-601.

⑬ See D. P. O'Connell, supra. note 29, at 10.

⑭ See Lesther Antonio Ortega Lemus, *Brief Outline of the History and Development of the Law of the Sea*, available at: http://www.academia.edu/1193093/Brief_Outline_of_the_History_and_Development_of_the_Law_of_the_Sea (last visited on 23 July 2013).

⑮ Ibid.

⑯ See Bernard G. Heinzen, supra. note 30, at 601.

I. Introduction

Grotius argued in favour of the liberty of the seas and oceans with a few exceptions: inlets, gulfs, inner seas, straits and what he defined as "all the expanse of sea which is visible from the shore". [47] It can be seen that under the Grotius' doctrine of "freedom of the seas", national rights and jurisdiction could be exercised over an undefined narrow belt of sea surrounding a nation's coastline. While Grotius firstly proposed a choice of the reach of human eye for the limits of coastal State authority, later on he propounded that the width of that strip of sea should be restricted to the capacity of the coastal State to exercise an effective control over it. [48] In order to translate Grotius' rule into practical terms, a Dutch writer stated that such control necessarily depends on the effective range of the weapons of the kingdom. [49] Although uncontested, most scholars attribute the determination of the range of the cannon shot in those times at three nautical miles or a league. [50]

It is argued that "[i]n theory, the Elizabethan position extended the *res communis* of the high seas to the shore of every coastal state"[51]. Elizabeth, however, was willing to recognize that "in some small distance from the coast" there may be "some oversight and jurisdiction". [52] Consequently, it can be seen that the Elizabethan and Dutch opinions coincided insofar as they were willing to recognize some supervisory jurisdiction within cannon range from the shore. [53]

The philosophy of effective power asserted over a narrow belt of sea surrounding a nation's coastline was subject to an intermittent stage of controversy and ambiguity regarding which claims to various forms of

---

[47] See Hugo Grotius, *The Freedom of the Seas* (Oxford University Press, 1916), at 37. Excerpted from Lesther Antonio Ortega Lemus, supra. note 44.
[48] See Lesther Antonio Ortega Lemus, supra. note 44.
[49] Ibid.
[50] Ibid. More detailed analysis sees Bernard G. Heinzen, supra. note 30.
[51] See Bernard G. Heinzen, supra. note 30, at 601.
[52] Ibid.
[53] Ibid.

jurisdiction could be made and to what distance of such power could be enforced.[54] These questions were not settled before the modern territorial sea and three-mile limit was ultimately accepted by States as customary international law.[55]

The stability attained in the law of the sea in the second half of the nineteenth century warranted the codification of its rules.[56] During the 1880s and 1890s, the Institute de Droit International devoted itself to establish the geographical areas for the choice of maritime laws, while it ran into difficulties on the questions both of the extent and the nature of territorial waters.[57] In 1930 the League of Nations, which was an intergovernmental organization founded in 1919 with the mission to maintain world peace, called for a conference in the Hague to codify, among other things, the law relating to territorial waters. Yet, due to the strong fragmentation of the participating States' position, the Hague Conference failed to reach an agreement with regards to the width of the territorial sea.[58]

After World War II, an impetus to extend national claims emerged. In 1945, the United States issued two statements. The first statement concerned the natural resources of the subsoil and seabed of the continental shelf, asserting the United States' exclusive right to exploit the mineral resources of its continental shelf beyond its territorial sea.[59] The other statement was relevant to the coastal fishery resources in certain areas of the high seas, claiming the exclusive conservation authority within a designated

---

[54] See D. P. O'Connell, supra. note 29, at 18.
[55] See Bernard G. Heinzen, supra. note 30, at 602.
[56] See D. P. O'Connell, supra. note 29, at 20.
[57] Ibid.
[58] See Lesther Antonio Ortega Lemus, supra. note 44.
[59] See Myron H. Nordquist (ed.), United Nations Convention on the Law of the Sea 1982: A Commentary, Vol. I (Martinus Nijhoff Publishers, 1989), at 1. The document is available at: http://www.ibiblio.org/pha/policy/1945/450928a.html (last visited on 14 April 2013).

I. Introduction

zone called "conservation zones".[60] This action served as a precedent for other States that had special interests in ocean resources, especially Latin American States, who proclaimed further rights over the vast ocean.

With the establishment of the International Law Commission of the United Nations (ILC), an opportunity was immediately presented to start working on treaty-making regarding the law of the sea.[61] The history of the modern law of the sea has remained a history of balancing conflicting interests between maritime and coastal States.[62] The primary reasons, which drove the rapid developments of the law of the sea, were the technological advances and the hunt for resources,[63] which meant that States' interests in the oceans were not just confined to navigation and fishing anymore.[64] These advanced technologies made it possible to exploit both living or non-living natural resources off their coasts. The international law of the sea was therefore intended to set out the principles and framework for regulating human's behaviour, as well as managing the resources of the oceans and the environment. A feature of the international law of the sea, however, is that it has been in an on-going state of development based on State practice, the views of publicists, or via new international treaties and instruments.[65] Notwithstanding its century-old history, it was and still is a contemporary product. Certain issues were unknown or not considered relevant at the time when the LOS Convention and its predecessors were negotiated. This poses

---

[60] See Myron H. Nordquist (ed.), United Nations Convention on the Law of the Sea 1982: A Commentary, Vol. I (Martinus Nijhoff Publishers, 1989), at 1. The document is available at: http://www.ibiblio.org/pha/policy/1945/450928a.html (last visited on 14 April 2013).

[61] See Donald R. Rothwell, Tim Stephens, supra. note 37, at 6.

[62] See Ted L. McDorman, Alexander J. Bolla, Douglas M. Johnston and John Duff, *International Ocean Law: Materials and Commentaries* (Carolina Academic Press, 2005), at 16.

[63] See Bo Johnson Theutenberg, *The Evolution of the Law of the Sea* (Tycooly International Publishing Limited, 1984), at 2.

[64] See Ted L. McDorman, Alexander J. Bolla et al., supra. note 62.

[65] See Donald R. Rothwell, Tim Stephens, supra. note 37, at 25.

potential problems as the law is increasingly under pressure to comprehensively manage the use and protection of the oceans.[66]

The United Nations convened three separate conferences on the law of the sea with the hope that guidelines and rules could be established for regulating all activities concerning the oceans.[67] The first one took place in 1958 and resulted in the four Geneva Conventions on the Law of the Sea: the Convention on the Territorial Sea and the Contiguous Zone;[68] the Convention on the High Seas;[69] the Convention on Fishing and Conservation of the Living Resources of the High Seas;[70] and the Convention on the Continental Shelf.[71] The second convention took place in 1960, which Thomas A. Clingan described as having "[...] the sole agenda of seeking agreement on the legal breadth of the territorial sea [but was] an endeavor that failed"[72]. Ted L. McDorman argued that the 1958 conferences were to be considered as failures as well, saying that:

> The 1958 and 1960 LOS Conferences did little to inhibit the coastal [S]tates from expanding their jurisdictional claims and, in fact helped the trend towards coastal State expansion by their failure to resolve three key issues: i) the width of the territorial

---

[66] See Donald R. Rothwell, Tim Stephens, supra. note 37, at 25.

[67] See Thomas A. Clingan, *The Law of the Sea: Ocean Law and Policy* (Austin&Winfield San Francisco-London, 1994), at 1. Also see Ted L. McDorman, Alexander J. Bolla et al., supra. note 62, at 17.

[68] United Nations Conference on the Law of the Sea, Official Records, Volume II: Plenary Meetings, A/CONF. 13/L. 52, at 132.

[69] United Nations Conference on the Law of the Sea, Official Records, Volume II: Plenary Meetings, A/CONF. 13/L. 53, at 135.

[70] United Nations Conference on the Law of the Sea, Official Records, Volume II: Plenary Meetings, A/CONF. 13/L. 54, at 139.

[71] United Nations Conference on the Law of the Sea, Official Records, Volume II: Plenary Meetings, A/CONF. 13/L. 55, at 142.

[72] See Thomas A. Clingan, supra. note 67, Introduction to the Law of the Sea. The same view is also put forward by Ted L. McDorman, Alexander J. Bolla et al., supra. note 62, at 18-19.

I. Introduction

sea; ii) the width of the continental shelf; and iii) the extent of coastal State control over fishing. [73]

After many years of intensive preparatory work, the third conference began in 1973. Compared to the earlier conferences, the most obvious feature was that lots of the acceding States were newly independent and developing countries,[74] due to the collapse of colonialism around the world. These new participants, whose concerns and interests were not the same as those of the maritime countries, held different political inclinations and infused new blood into the deliberations. It turned out that these countries had an immense impact on shaping the contents of the later LOS Convention.[75] Frequently, the LOS Convention is referred to as the "constitution for the oceans", due to the fact that it divided the oceans both horizontally and vertically.[76] From the horizontal viewpoint, the oceans were divided into the territorial seas, contiguous zones, exclusive economic zones (EEZ), and the high seas. Vertically, the marine areas were divided into the airspace above the ocean, the water column, seabed, ocean floor, and subsoil below. In the territorial sea, coastal States enjoy sovereignty that extends to the airspace over their territorial sea as well as to its seabed and subsoil.[77] The maximum breadth of territorial sea was finally fixed at twelve nautical miles.[78] All ships enjoy the right of innocent passage through the territorial sea.[79] The contiguous zone is the marine area along the territorial sea, which does not extend beyond twenty-four nautical miles from the baselines.[80] Within this area, coastal States have the right to adhere to their customs, fiscal, immigration or sanitary laws, and regulations valid within

---

[73] See Ted L. McDorman, Alexander J. Bolla et al., supra. note 62, at 18-19.
[74] See Thomas A. Clingan, supra. note 67, Introduction to the Law of the Sea.
[75] Ibid.
[76] Ibid.
[77] LOS Convention, Article 2.
[78] LOS Convention, Article 3.
[79] LOS Convention, Article 17.
[80] LOS Convention, Article 33.

their territory or territorial sea from being infringed. The contiguous zone can be considered as a transition zone, as far as its status is concerned, located "between" the territorial sea (sovereignty) and the high seas (freedom).[81] In the EEZ, which may extend to 200 nautical miles from the baselines,[82] the LOS Convention provides coastal States with functionally limited sovereign rights ("constituting an extract of the comprehensive concept of sovereignty")[83] and jurisdiction. Alexander Proelss has argued that "[t]hese rights are not associated with the zone in a spatial sense, but mainly derive from its economic potential"[84]—a conclusion that is supported by LOS Convention Article 56 (1) (a) ("for the purpose of exploring, exploiting, conserving and managing the natural resources, whether living or non-living").[85] The legal regime on the continental shelf, which comprises the seabed and subsoil of the submarine areas,[86] is complicated, particularly because of the difficulties in determining the continental margin. The LOS Convention lays down specific criteria and procedures concerning the ascertainment of the margin's outer limit.[87] According to the LOS Convention, coastal States enjoy sovereign rights over the exploration and exploitation of non-living resources, including oil and gas, found in the seabed and subsoil of the continental shelf.[88] The LOS Convention also establishes an international regime to regulate the exploration and exploitation of the seabed's non-living resources beyond the limits of national

---

[81] See Thomas A. Clingan, supra. note 67.
[82] LOS Convention, Article 57.
[83] See Alexander Proelss, The Law on the Exclusive Economic Zone in Perspective: Legal Status and Resolution of User Conflicts Revisited, in: *Ocean Yearbook*, Vol. 26, 2012, 87-112, at 93.
[84] Ibid., at 89.
[85] LOS Convention, Article 56.
[86] LOS Convention, Article 76.
[87] See Ted L. McDorman, Alexander J. Bolla et al., supra. note 62, at 28. For details see LOS Convention, Article 76.
[88] LOS Convention, Article 77.

◇ I . Introduction ◇

jurisdiction.[89] The International Seabed Authority is charged with managing or overseeing any intended exploration and exploitation. Regarding the protection of the marine environment, the LOS Convention sets forth some basic obligations and a comprehensive legal framework to prevent the marine environment from different kinds of pollution.[90]

C. Problem Identification

According to Ted L. McDorman, "[m]arine scientific research serves a wide variety of purpose"[91]. First and most importantly, exploration and exploiting vast ocean resources depends on thorough MSR. Second, the studies of waves, currents and the weather also serve shipping and navigation. Third, actualizing interests that lie in the protection of the marine environment and the conservation of its resources primarily depend on developing MSR. Furthermore, MSR can increase marine knowledge for the benefit of all human.[92]

Naturalists on land, not surprisingly, can travel the whole world in search of knowledge.[93] Geological research activities do not appear to be hindered by national sovereignty.[94] More than a century ago, oceanographers could also travel the oceans with the absence of regulation and protocol as

---

[89] Relevant articles see LOS Convention, Part XI.

[90] See Ted L. McDorman, Alexander J. Bolla et al., supra. note 62, at 29. For details see LOS Convention, Part XII.

[91] See R. R. Churchill, A. V. Lowe, *The Law of the Sea* (Manchester University Press, 1999), at 400.

[92] Ibid.

[93] See John A. Knauss, Development of the Freedom of Scientific Research Issue of the Third Law of the Sea Conference, in: *Ocean Development and International Law Journal*, Vol.1, 1973, 93-120, at 94.

[94] Ibid.

naturalists on land did.[95] Viewed from this perspective, MSR has undergone a transition from being free of control to becoming increasingly regulated. In the past, scientists autonomously planned their cruises, made arrangements and conducted the research wherever they wanted. They may have contacted officials in the nations off whose coasts they worked, but they often worked without officially informing the coastal State of their presence.[96] Even if there were restrictions, a possible way to alleviate that effect was to contact research colleagues in the States concerned, inviting them to come on the cruise.[97] Since World War Ⅱ, with the shrinking of the high seas, the new and intensified uses of the oceans and its resources have made MSR more important due to its practical application to natural resource exploitation, marine environmental protection and military purposes. None of these purposes are possible without greatly improved understanding of how the ocean works.[98] Due to the significance of MSR, together with its past successes, "science and scientists are no longer considered as harmless or as innocent as was the case a century ago".[99] MSR is regarded as a double-

---

[95] See John A. Knauss, Development of the Freedom of Scientific Research Issue of the Third Law of the Sea Conference, in: *Ocean Development and International Law Journal*, Vol. 1, 1973, 93-120, at 94.

[96] See Judith A. Tegger Kildow, Nature of the Present Restrictions on Oceanic Research, in Warren S. Wooster (ed.), *Freedom of Oceanic Research: A Study Conducted by the Center for Marine Affairs of the Scripps Institution of Oceanography* (Crane, Russak&Company, Inc, 1973), 5-28, at 9. The same view is taken by Herman T. Franssen, Developing Country Views of Sea Law and Marine Science, in Warren S. Wooster (ed.), *Freedom of Oceanic Research: A Study Conducted by the Center for Marine Affairs of the Scripps Institution of Oceanography* (Crane, Russak&Company, Inc, 1973), 133-177, at 150.

[97] See David A. Ross, *Opportunities and Uses of the Ocean* (Springer-Verlag New York, 1978), at 64.

[98] See Warren S. Wooster (ed.), *Freedom of Oceanic Research: A Study Conducted by the Center for Marine Affairs of the Scripps Institution of Oceanography* (Crane, Russak&Company, Inc, 1973), Introduction, at 2.

[99] See John A. Knauss, supra. note 93.

I. Introduction

edged sword, and it is never easy to say who is benefiting. The classic saying therefore is that "science's lack of success is perhaps as important as its success"[100]. Thus, a difference in views emerged between the old industrialized States who have money, technology and know-how to benefit from MSR, and the developing States who desired to protect their own interests. Developing countries sought and still seek to extend control over the conduct of MSR, including the discretion to refuse the permission.[101] In their opinion, all scientific research can be somewhat useful, at least indirectly, with respect to fisheries, oil or mineral exploitation and military operations. By contrast, industrialized countries, which can best utilize the research information, will immediately benefit from MSR. Nevertheless the research may indirectly benefit all of mankind.[102] For developing States, even if freedom is given to MSR, they do not have the economic, scientific and technological capacity to make use of that freedom. The alleged freedom to conduct MSR is thus somewhat illusory, and the equality before the law loses a portion of its reality.[103] It is well known that scientific knowledge has laid the foundations for exploration and exploitation of ocean resources. Developing countries fear that the ocean resources would be exploited mainly by the developed countries, partly because such countries would be able to better locate and identify the resources off their coasts or on their offshore seabeds. National security is another issue that worries developing States. Whenever groups from developed countries take measurements of water depth, temperature, pressure, or salinity, the collected data may have potential uses for their navies.[104] This issue is addressed by Herman T.

---

[100] See John A. Knauss, supra. note 95.

[101] See Patricia Birnie, Law of the Sea and Ocean Resources: Implications for Marine Scientific Research, in: *The International Journal of Marine and Coastal Law*, Vol. 10, 1995, 229-252, at 234.

[102] See Herman T. Franssen, supra. note 96, at 163.

[103] Ibid.

[104] Ibid.

Franssen, who stated that:

> [s]ome developing countries are reluctant to grant foreign research vessels access to their coastal waters because they fear that oceanographic vessels, with their sophisticated equipment, may engage in espionage or other illegal activities.[105]

It is often felt that scientists oftentimes have a non-political attitude about their research activities and simply focus on conducting their research for the sake of science in order to understand as much as they can about the ocean.[106] From a scientists' perspective, if the research activities are of some economic importance to the coastal State, the research effort should be welcomed rather than hindered.[107] The reason why marine scientists have expressed anxiety over the threat of increased restrictions on MSR is because they are already faced with the difficulties imposed by nature, and wish to avoid further restraints imposed by man.[108] However, many connections between MSR and other matters over which the international community wishes to exercise control—such as resource exploitation and military application of the ocean—have led the international community to regulate and restrict freedom to conduct MSR, first in the Geneva Conventions and latterly, and rather more extensively, in the LOS Convention.[109]

D. Research Questions

This book is concerned with MSR under the context of the LOS Convention, namely, the international rules and principles provided by the

---

[105] See Herman T. Franssen, supra. note 96, at 163.
[106] See John A. Knauss, supra. note 93, at 96.
[107] Ibid.
[108] Ibid.
[109] See R. R. Churchill and A. V. Lowe, supra. note 91, at 401. Also see John A. Knauss, supra. note 93.

Ⅰ. Introduction

LOS Convention that bind States in their international relations concerning marine scientific research. Accordingly, it does not, except incidentally, involve other legal regimes stipulated by the LOS Convention. Given that the LOS Convention was the result of a compromised package deal, several key issues regarding MSR were not clarified in order to achieve consensus, such as the definition of MSR, what comprises MSR (as far as the differentiation between pure and applied research contention is concerned), hydrographic surveying and so on. Under such circumstances, the legal regime provided for MSR by the LOS Convention is systematically analysed in this book. At the same time, particular focus is put on LOS Convention Article 246 regarding the international legal regime governing the conduct of MSR in coastal States' EEZ as well as on the continental shelf due to that it is one of the most prominent articles in the respect of regulating MSR. Various elements addressed in LOS Convention Article 246 represent a huge degree of complexity, and some of them are still subject to uncertainty. Moreover, it is the only article in the LOS Convention to address simultaneously the balancing of the substantive rights of the coastal State and other States both in the EEZ and on the continental shelf.[10]

Additionally, as the LOS Convention is now over a quarter of a century old, there are several contemporary oceans issues that are not adequately addressed by the LOS Convention. In the era of marine high-tech, the emergence of new (in part revolutionary) technologies has raised questions as to whether the assumptions and intentions of LOS Convention Part XIII[11] still hold true.[12] Ocean upwelling, voluntary observing ships, and marine bioprospecting, three particularly important and controversial examples of newly occurring ocean uses in the context of changes in marine science and

---

[10] See Myron H. Nordquist (ed.), *United Nations Convention on the Law of the Sea* 1982: *A Commentary*, Vol. Ⅳ (Martinus Nijhoff Publishers, 1989), at 517.

[11] LOS Convention Part XIII provides legal regime regarding marine scientific research.

[12] See Aldo Chircop, supra. note 27, at 575.

technology, are examined from the legal perspective. The factual backgrounds of these respective activities are initially brought forward in order to lay out a picture how these activities are conducted. The legal analysis, as the main component of the study, comprises the legal classification as well as jurisdictional issues of the three case studies. Furthermore, removal requirements (with respect to ocean upwelling pipes), the notifying regime (with respect to voluntary observing ships) and the applicability of the benefit sharing regime (with respect to marine bioprospecting) will be analysed. Through these case studies, the present analysis aims to contribute to a better understanding of the role of the International Law of the Sea as to newly emerging marine scientific research and related activities.

E. Structure of the book

The purpose of this book is to contribute to a better understanding of the international legal aspects involved in MSR. To this end, four chapters are organized in a systematic manner. Following this introduction, legal developments with respect to MSR under the International Law of the Sea are examined from a historical perspective. The analysis is then more focused on post-LOS Convention times. While an overall description is presented according to various kinds of State influence exerted on different maritime zones, Article 246 as one of the most prominent provisions in LOS Convention Part XIII,[13] is subject to a more detailed survey. The third chapter is a case study in terms of implementing the legal regime of MSR stipulated in the LOS Convention. Three newly emerging MSR related activities are

---

[13] As stated, LOS Convention Article 246 addresses the legal regime concerning MSR in the EEZ and on the continental shelf.

I. Introduction

analysed under the context of the LOS Convention. The fourth chapter then contains some final remarks that may shed light on the outlook of legal developments concerning MSR.

Ⅱ. Marine Scientific Research under the International Law of the Sea

There is no definition of the term "Marine Scientific Research" (MSR) in the LOS Convention. There was, however, a long struggle and controversial negotiation in defining the rights, obligations and duties of States in different maritime zones regarding conducting MSR. Awareness of the significance of MSR's numerous applications made its legal regime a central issue throughout the negotiating process at several Law of the Sea conferences.[114]

A. Prior to 1958

Before the first conference on the Law of the Sea in 1958, MSR had not gained too much attention and "was not considered as being among the major fields of maritime activity".[115]

The first focus on MSR was during the discussion about the legal

---

 [114] See Myron H. Nordquist (ed.), *United Nations Convention on the Law of the Sea 1982: A Commentary*, Vol. Ⅳ, supra. note 110, at 429.

 [115] Ibid. The same view is taken by R. R. Churchill and A. V. Lowe, supra. note 91, at 400, stating that "[u]ntil the middle of the twentieth century no legal controls on the conduct of marine scientific research were perceived to be necessary, and indeed the Law of the Sea literature up to this time contains virtually no mention of scientific research. This may be explained partly by the generally prevailing attitude that scientific research should be free of governmental controls, and partly by the modest scale and limited practical application of marine scientific research".

regime of the high seas at the seventh session of the International Law Commission (ILC) in 1955.[116] There were two drafts relating to the International Law of the Sea, adopted by the ILC during its seventh session, namely the "Provisional Articles Concerning the Regime of the High Seas" and the "Draft Articles on the Regime of the Territorial Sea".[117] During the 320th meeting of the seventh session, France insisted on retaining the words "inter alia" in Article 2 of the "Provisional Articles Concerning the Regime of the High Seas", which listed the freedoms of the high seas since there were other freedoms covered by Article 2, such as the right of conducting scientific research.[118] Finally, the words "inter alia" were retained in the Commission's draft Article 2 included in the ILC report of its seventh session. The Commission then invited the governments to submit their observations on these drafts.[119] By July 2nd, 1956, 25 members of United Nations had submitted their comments.[120] Only one of them made reference to the issue of MSR: the United Kingdom (UK).[121] The UK proposed adding

---

[116] See A. H. A. Soons, *Marine Scientific Research and the Law of the Sea* (The Hague, 1982), at 49. The ILC took the view that it should set aside a number of subjects the limited importance of which did not appear to justify their consideration in the present phase of its work.

[117] See International Law Commission Seventh Session, 1955, A/CONF. 4/94, available at: http://untreaty. un. org/ilc/documentation/english/a _ cn4 _ 94. pdf (last visited on 14 April 2013). It should be pointed out that, at ILC's seventh session, the Commission studied the subject on the basis of the sixth session report and adopted a provisional draft and comments which are reproduced. The draft articles of the sixth session see International Law Commission Sixth Session, 1954, A/CONF. 4/79.

[118] See Yearbook of the International Law Commission 1955, Vol. I, p. 222. http://untreaty. un. org/ilc/publications/yearbooks/Ybkvolumes(e)/ILC _ 1955 _ v1 _ e. pdf (last visited on 14 April 2013).

[119] See International Law Commission Eighth Session, 1956, A/CONF. 4/99, http://untreaty. un. org/ilc/documentation/english/a_cn4_99_add1-9. pdf (last visited on 14 April 2013).

[120] Ibid.

[121] More detailed information about the comment sees Ibid.

a fifth item to the freedoms in Article 2 of the regime of the high seas, which was "freedom of research, experiment and exploration". [122] The reason put forward for this suggestion by the UK government was that it had received evidence that a number of learned and scientific bodies were concerned that recent developments would impede the freedoms of research, exploration and experiment. [123] It was finally agreed that the fifth item with respect to the freedom of scientific research in the high seas would not be included in the text of the Article, but that it would be retained in the commentary thereto. [124] Besides, the UK government also made extra comments on the draft articles of the continental shelf and the contiguous zone, which were drawn up at the fifth session of the International Law Commission in 1953. [125] In the comment relating to the continental shelf, the UK commented that:

> [...] certain scientific societies are concerned lest the terms of these Articles should enable the coastal State to place unnecessary restrictions upon *bona fide* scientific research upon the shelf itself. They therefore suggest that the Commission consider inserting some provisions safeguarding the general right

---

[122] More detailed information about the comment sees Ibid.

[123] Ibid. Also see A. H. A. Soons, supra. note 116, at 50.

[124] See A. H. A. Soons, Ibid., at 117. The document sees Yearbook of the International Law Commission 1956, Vol. I, at 32. http://untreaty.un.org/ilc/publications/yearbooks/Ybkvolumes(e)/ILC_1956_v1_e.pdf (last visited on 14 April 2013). Here it should be pointed that there is an ambiguity that which Commentary here referred to. Whether the Commentary accompanied with the Provisional Articles concerning the Regime of the High Seas which adopted in July 1955 or the Commentary connected with the Articles concerning the Law of the Sea adopted in July 1956 which will be introduced later. According to the time when the Government Comment of UK Government was offered was in 1956, it would be seen that here the Commentary means the latter one.

[125] More detailed comments see supra. note 119, at 86.

II. Marine Scientific Research under the International Law of the Sea

to undertake such exploration and research.[126]

The above comments influenced the discussion as well at the 359th meeting, during the ILC's eighth session.[127] Finally, members of the Commission agreed to include a reference to MSR in the Commentary instead of having an express article provision for it.[128]

Later, pursuant to General Assembly Resolution 899 (IX) of December 14th, 1954,[129] the Commission, at its eighth session in 1956, grouped systematically together all the rules that it had adopted concerning the high seas, the territorial sea, the continental shelf, the contiguous zone and the conservation of the living resources of the sea.[130] These are known as the "Articles Concerning the Law of the Sea".[131] The articles' sequence was rearranged, with the first part dealing with the territorial sea and the second one dealing with the high seas. The second part was divided into three sections: (1) general regime of the high seas; (2) contiguous zone; (3) continental shelf.[132] Each article was provided with a Commentary section.[133] Among the whole set of articles, only the Commentaries as regards to the freedoms of

---

[126] More detailed comments see supra. note 119, p. 87. Also see A. H. A. Soons, supra. note 116, at 58.

[127] The detailed discussions see Year Book of International Law Commission, Vol. I, 1956, at 147. http://untreaty.un.org/ilc/publications/yearbooks/Ybkvolumes (e)/ILC_1956_v1_e.pdf (last visited on 14 April 2013).

[128] Ibid., at 147-148. Also see A. H. A Soons, supra. note 116, at 58-59. The Commentary about it will be analysed later.

[129] The document sees http://daccess-dds-ny.un.org/doc/RESOLUTION/GEN/NR0/096/34/IMG/NR009634.pdf? Open Element (last visited on 14 April 2013).

[130] See International Law Commission Eighth Session, 1956, A/CONF. 4/104. http://untreaty.un.org/ilc/documentation/english/a_cn4_104.pdf (last visited on 14 April 2013).

[131] Ibid.

[132] Ibid.

[133] Ibid. Or more detailed information about the Commentary sees http://untreaty.un.org/ilc/texts/instruments/english/commentaries/8_1_8_2_1956.pdf (last visited on 14 April 2013).

the high seas and the continental shelf, particularly those referring to Articles 27 and 68, made some references to MSR.[134] In the Commentary to Article 27, which was about the freedom of the high seas, it expressed that:

[t]he list of freedoms of the high seas contained in this article is not restrictive. The Commission has merely specified four of the main freedoms, but it is aware that there are other freedoms, such as freedom to undertake scientific research on the high seas. [sic][135]

In the meantime, Article 68 addressed the sovereign rights of coastal States on the continental shelf. There was also a contested discussion with respect to the legal regime of scientific research on the continental shelf at the 359th meeting of the eighth session.[136] The Commentary relating to Article 68 pointed out several issues, such as adopting more accurate terms and the details on implementation.[137] Among other issues, it was decided that although the concept of the continental shelf did not encompass the international submarine areas, there was still a possibility that an international body for scientific research could be set up in the future under the framework of an existing international organization.[138]

The other important discussion point related to the anxiety within scientific circles. They were concerned that the "freedom to conduct scientific research in the soil of the continental shelf and in the waters above

---

[134] See Myron H. Nordquist (ed.), *United Nations Convention on the Law of the Sea* 1982: *A Commentary*, Vol. Ⅳ, supra. note 110, at 429-430.

[135] See International Law Commission Eighth Session, 1956, A/CONF. 4/104, supra. note 130, at 278. And also see Myron H. Nordquist, ibid, at 430.

[136] See Yearbook of International Law Commission, Vol. Ⅰ, 1956, at 147. http://untreaty.un.org/ilc/publications/yearbooks/Ybkvolumes(e)/ILC_1956_v1_e.pdf (last visited on 14 April 2013). Also see A. H. A. Soons, supra. note 116, at 58.

[137] The Commentary on article 68 sees International Law Commission Eighth Session,1956, A/CONF.4/104,supra. note 130, at 297.

[138] Ibid., at 298.

would be endangered"[⑬]. In the Commentary, it was assured that when research was conducted in the waters above the continental shelf, freedom would be "in no way affected", since the water column would still comprise the high seas. [⑭] In the meantime, coastal States "will not have the right to prohibit the scientific research" if they had no relationship with the exploration or exploitation of the seabed or subsoil. [⑮] However, it should be taken into account that the freedom of scientific research as referred to in the Commentary only mentioned "the waters above", whereas the legal situation remained vague with regard to scientific research "in the soil of the continental shelf". [⑯] However, with the development of marine science, the vagueness concerning scientific research in the soil of the continental shelf was addressed during later negotiations.

B. 1958 the First UN Conference on the Law of the Sea (UNCLOS Ⅰ)

From February 24th to April 27th, 1958, the United Nations Conference on the Law of the Sea (UNCLOS I) was held at the European Office of the United Nations at Geneva. [⑰] The draft articles formulated by

---

　　[⑬] The Commentary on article 68 sees International Law Commission Eighth Session, 1956, A/CONF. 4/104, supra. note 130, at 298.
　　[⑭] Ibid.
　　[⑮] Ibid.
　　[⑯] Ibid. It was expressly stated that "it was thought that freedom to conduct scientific research in the soil of the continental shelf and in the waters above would be endangered".
　　[⑰] See http://untreaty.un.org/cod/avl/ha/gclos/gclos.html (last visited on 14 April 2013).

the ILC formed the basis of the conference discussions.[144] Four Conventions and a Protocol[145] were adopted at the conference and were subsequently opened for signature.

1. MSR in the High Seas

Attended by 86 States, the conference was organized into five main committees and a plenary. During the discussion, the Second Committee was responsible for the general regime of the high seas. From the fourth to thirty-fourth meeting of the Second Committee, the subject matter centred on considering the draft articles[146] adopted by the ILC at its eighth session. At the fifth meeting, Lebanon expressed that "[a]dditional clauses might be added to Article 27 providing for freedom of scientific research and exploration and other kinds of freedom mentioned in the commentary of the International Law Commission".[147] Likewise, at the fifteenth meeting, Portugal proposed that the fifth freedom, the "freedom to undertake research, experiments and exploration", should be mentioned in Article 27.[148] There then had been silence on the issue for a while until the twenty-second meeting. The Mexican delegation expressed its objection against Portugal's proposal on the issue of the fifth freedom without giving any reasons and made no comment relating to it.[149] However, the Spanish

---

[144] See A. H. A. Soons, supra. note 116, at 52.

[145] The conventions are Convention on the Territorial Sea and the Contiguous Zone (A/CONF. 13/L. 52); Convention on the High Seas (A/CONF. 13/L. 53); Convention on Fishing and Conservation of the Living Resources of the High Seas (A/CONF. 13/L. 54); Convention on the Continental Shelf (A/CONF. 13/L. 55). And the protocol is Optional Protocol of Signature concerning the compulsory settlement of disputes (A/CONF. 13/L. 57).

[146] This term refers to the "Articles concerning the Law of the Sea" adopted by ILC.

[147] See United Nations Conference on the Law of the Sea, Official Records, Vol. IV: Second Committee, Summary Records of Meetings and Annexes, at 5.

[148] Ibid., at 37-38.

[149] Ibid., at 55.

delegate later supported Portugal's proposal.⑮⁰ In the subsequent voting, Portugal's proposal was rejected without any further discussion by 39 votes to 13, with 8 abstentions.⑮¹

The high seas freedoms in the Convention on the High Seas 1958 (CHS 1958) are essentially the same as those contained in the draft articles.⑮² However, an additional paragraph was inserted, and some of the words made the freedoms even more obscure: "[t]hese freedoms, and others which are recognized by the general principles of international law, shall be exercised [...]".⑮³ There was no guidance in CHS 1958 Article 2 on how to interpret the phrase "others which are recognized by the general principles of international law", neither as to the meaning of "general principles of international law".⑮⁴ As such, since it would evoke diverse comprehensions with regards to the question on how to conduct MSR, is it to be regarded as an

---

⑮⁰ See United Nations Conference on the Law of the Sea, Official Records, Vol. Ⅳ: Second Committee, Summary Records of Meetings and Annexes, at 55.

⑮¹ Ibid.

⑮² See Convention on the High Seas (A/CONF. 13/L. 53). The contents of freedoms in high seas were showed "... It comprises, inter alia, both for coastal and non-coastal States: (1) Freedom of navigation; (2) Freedom of fishing; (3) Freedom to lay submarine cables and pipelines; (4) Freedom to fly over the high seas".

⑮³ See Convention on the High Seas. Ibid.

⑮⁴ See A. H. A. Soons, supra. note 116, at 53.

exercise of the freedom of the high seas or not?[148]

2. MSR on the Continental Shelf

At the conference, the legal regime of the continental shelf was dealt with in the Fourth Committee and ample attention was given to the issue of MSR.[149] Therefore, the restrictions on MSR are mainly embodied in the Convention on the Continental Shelf 1958 (CCS 1958). Article 5 (1) stated that:

[...] **exploration of the continental shelf and the exploitation of its natural resources must not [...] result in any interference with fundamental oceanographic or other scientific research carried out with the intention of open publication.**[150]

This is a general principle regarding protection of MSR on the continental shelf. Further, it also should be pointed out that Article 5 (1) of the CCS 1958 only makes reference to the "fundamental oceanographic or other scientific research carried out with the intention of open publication".

---

[148] See A. H. A. Soons, supra. note 116, at 53. Soons proposed to analyse how to understand the last sentence in Article 2. If it is understood that the freedom of conducting marine scientific research should be expressly recognized by international law, the "activity can probably only become a recognized 'freedom of the high seas' by means of the process of the formation of customary international law". Under this approach, in Soons' view, the conduct of marine scientific research by vessels "definitely qualifies as a freedom of the high seas, since this activity has been carried out for over a century without any protests, and by a large number of States". If it is understood that, another approach, "any use of the high seas is permitted, provided it does not interfere unreasonably with other uses". In Soons' view, under such circumstances, "marine scientific research can be conducted freely in the high seas, subject only to the requirement of reasonable regard to the interests of the other uses of the high seas". He also made a list of the scholars on the Law of the Sea who interpreted the last sentence of Article 2, like Brown, and Burke, and some other different opinions as well, like Dixit and Menzel. The exact articles of all the authors see Soons' book notes 60-71 of Part Two.

[149] Ibid. at 61.

[150] See Convention on the Continental Shelf (A/CONF. 13/L. 55).

Therefore, there would be significant difficulties arising from the wording.[158] Moreover, Article 5 (8) of the CCS 1958 contained a rule that was sufficient to arouse concerns among scientific circles.[159] It stated that:

> **The consent of the coastal State shall be obtained in respect of any research concerning the continental shelf and undertaken there. Nevertheless, the coastal State shall not normally withhold its consent if the request is submitted by a qualified institution with a view to purely scientific research into the physical or biological characteristics of the continental shelf, subject to the proviso that the coastal State shall have the right, if it so desires, to participate or to be represented in the research, and that in any event the result shall be published.**[160]

While the general principle and idea can be easily grasped, the details of the drafting reveal many possible ambiguities, as discussed below.

(a) Fundamental, Pure Research versus Applied Research

First, according to Article 5 (1) of the CCS 1958, one can deduce that the exploration and exploitation of the natural resources are different from the "fundamental oceanographic or other scientific research carried out with

---

[158] More details see, Michael Redfield, The Legal Framework for Oceanic Research, in Warren S. Wooster (ed.), *Freedom of Oceanic Research: A Study Conducted by the Center for Marine Affairs of the Scripps Institution of Oceanography* (Crane, Russak&Company, Inc., 1973), 45-95, at 53, concludes that "What is 'fundamental oceanographic or other scientific research'? Who is to define it? Who is to apply the definition to a given project or case? What objective criteria, if any, establish the requisite 'intention of open publication'? Who determines if this intention is present? And, lastly, what is the status of research which does not qualify as 'fundamental', or which is not to be carried out 'with the intention of open publication'?".

[159] More details see, Judith A. Tegger Kildow, supra. note 96, at 5. It is noted that "Immediately following the ratification of the Geneva Convention on the Continental Shelf, the freedom of access problems for scientific research was discussed with some concern in the sessions of the two intergovernmental organizations principally dedicated to facilitating oceanographic research".

[160] See Convention on the Continental Shelf (A/CONF. 13/L. 55).

the intention of open publication". Second, based on Article 5 (8) of the CCS 1958, "purely scientific research into the physical or biological characteristics of the continental shelf" is not the only kind of research on the continental shelf.

Contrary to Article 5 (1) of the CCS 1958, Article 5 (8) of the CCS 1958 is aimed at "any research". It means that with regard to both pure and applied research, the coastal State's consent should be obtained.[16] Under normal circumstances, however, pure research is not supposed to be denied if various conditions are complied with.[16] In this regard, the two provisions, Article 5 (1) and Article 5 (8) of the CCS 1958, seem to conflict each other. P. K. Mukherjee expressed that:

> [W]hile paragraph 1 seemed to reflect an inclination towards preserving a general freedom of research by drawing an apparent distinction between fundamental and applied research, paragraph 8, by superimposing a proviso and despite the distinction, seemed to have created some equivocality.[18]

Regarding the phrases "pure and applied research", "fundamental research" or "resource oriented", and other similar phrases, their definitions and the distinction between them remained vague at that time. The CCS 1958 intended to distinguish them, rather than defining them, by formulating different reactions and consequences due to the types of the research. However, it should be pointed out that "applied" research covers a broad range of research. The notion that the research is of "direct significance for the exploration and exploitation of natural resources" derives from the concept of "applied research". For the coastal State, the research that is of "direct significance for the exploration and exploitation of natural resources" is a matter of serious concern. Pursuant to LOS Convention

---

[16] See R. R. Churchill and A. V. Lowe, supra. note 91, at 402.

[16] Ibid., at 402.

[18] P. K. Mukherjee, The Consent Regime of Oceanic Research in the New Law of the Sea, in: *Marine Policy*, Vol. 5, 1981, 98-113, at 101.

Article 246 (5) (a), this is also one of the circumstances where coastal States can withhold their consent.

Other kinds of MSR may have to be considered as "applied" in a functional sense, but are undoubtedly significant for the benefit of all the human beings,[164] such as hydrographic surveying, whose objective is the safety of navigation. Without hydrographic surveying, the safety of international shipping, trade and ocean navigation would be at stake. The charting of reefs and other hidden dangers at sea are also critical to safe navigation.[165] With regard to other situations, Soons stated that:

> **Examples of such research[applied research] are chemical oceanographic investigations conducted for the purpose of marine pollution control measures, physical oceanographic investigations carried out for the purpose of enhancing long-range weather forecasting or for military purpose (e. g., anti-submarine warfare), and marine biological investigations for the purpose of managing living marine resources.[166]**

Thus, it should be borne in mind that the position towards this kind of "applied" research should be different from the "resource-oriented" research that is also part of "applied" research. The controversy concerning the concept's scope of MSR is mostly centred on the difference between "resource-oriented" research and "pure" research.

(b) Research Concerning the Continental Shelf and Undertaken There

Another ambiguous provision, "research concerning the continental

---

[164] P. K. Mukherjee, The Consent Regime of Oceanic Research in the New Law of the Sea, in: *Marine Policy*, Vol. 5, 1981, 98-113, at 113.

[165] Ibid.

[166] See A. H. A. Soons, supra. note 116, at 7.

shelf and undertaken there", could lead to several interpretations.[167] Among these interpretations, the decisive factor is how to understand which kind of research is in question. Two different approaches could be taken. On one hand, it could refer to research concerning the continental shelf and that is physically undertaken on the shelf. On the other hand, it could cover research concerning the continental shelf—regardless of it being carried out on the continental shelf or in the superjacent water—and the research which is conducted on the continental shelf, whether it concerns the continental shelf or not.[168] However, there was no consensus achieved as to what method of interpretation is correct, and the CCS 1958 gave no guidance as well.[169] Due to the somewhat unclear wording, authors held varied opinions regarding how to interpret "research on the continental shelf", which is subject to the consent requirement. R. R. Churchill and A. V. Lowe addressed this issue in the following terms:

> **As to which interpretation is correct, the travaux preparatoires of Continental Shelf Convention give little guidance, while the practice of parties to the Convention, though not conclusive, appears to support the second interpretation.**[170]

Even as far as the second interpretation is concerned, disagreements still exist. Soons stated that:

> It is therefore submitted that paragraph 8 should be interpreted as requiring the consent of the coastal State for all scientific research, whatever the method used, concerning the

---

[167] For a deeper analysis see R. R. Churchill and A. V. Lowe, supra. note 91, at 402. Also see Michael Redfield, supra. note 158, at 54. It is stating that "the term 'research concerning the continental shelf and undertaken there' is not uniformly interpreted by states, and some interpretations conflict with those drawn above, with regard to research undertaken in the water column over the shelf".

[168] See R. R. Churchill and A. V. Lowe, supra. note 91, at 402.

[169] Ibid.

[170] Ibid.

continental shelf.[171]

In his view, any research "concerning the continental shelf" is supposed to require the prior consent of the coastal State without considering whether it is physically undertaken there or not. He convincingly concluded that:

> This view leads to the conclusion that the words "and undertaken there" in the first sentence of paragraph 8 have no meaning: they do not mean that the scientific research activities covered by it must involve physical contact with the continental shelf, nor do they mean that any research involving physical contact with the shelf is covered.[172]

He also made a further observation by offering two circumstances to justify his conclusion. The first is that certain investigations are concerned with the natural resources of the continental shelf instead of the continental shelf itself. Furthermore, as stated above and according to Soons' perspective, since the research cannot be regarded as concerning the continental shelf, it is submitted that the coastal States, in such a situation, are not in the position where their consent is required.[173] The second issue is that of research concerning the superjacent waters conducted by deploying devices that rest on the seabed, such as instrument packages, fixed installations or anchored buoys.[174] This is a perfect instance of the kind of research that does not concern the continental shelf or its nature resources but involves physical contact with the continental shelf. Soons explained that: "unless the deployment of such devices, in a particular situation, could interfere with the exercise by the coastal State of its sovereign rights to explore and exploit the continental shelf, the conduct of the scientific research does not need the prior consent of the coastal State."[175] Following

---

[171] See A. H. A. Soons, supra. note 116, at 71.
[172] See Michael Redfeld, supra. note 158, at 72.
[173] Ibid., at 71-72.
[174] Ibid., at 72.
[175] Ibid., at 72.

this consideration, a crucial factor that should not be ignored is the period of time for which the devices are deployed on the seabed. Temporarily deploying devices are unlikely to interfere with the coastal State's rights, while a totally different situation emerges if the construction is a (semi-) permanent platform, whether manned or not.⑯ Despite the fact that according to the CCS 1958, any State, coastal or not, may deploy installations or devices on the coastal State's continental shelf for peaceful purposes other than exploring or exploiting natural resources,⑰ a (semi-) permanent platform that is normally constructed with concrete or steel fixed to the sea floor with piles would very much likely constitute an unreasonable interference with the coastal State's right to explore and exploit the natural resources on the continental shelf. Furthermore, pursuant to Soons, the coastal State may still exercise a large degree of discretion in determining whether or not a device constitutes an "unreasonable" interference since the interpretation of "interference" may include possible future interferences even if there is no exploitation activity in the proposed area yet.⑱ Under this circumstance, the deployment of a (semi-) permanent platform used for conducting MSR concerning the superjacent water is still subject to the coastal State's consent.⑲

Contrarily, Michael Redfield addressed this issue arguing that: "[o]ther evidence, however, indicates that physical contact is regarded as necessary by the State Department before research comes under coastal control as shelf research."⑳ No reference was made to the phrase "concerning the continental shelf". Viewed from this perspective, Redfield

---

⑯ See Michael Redfield, supra. at Note to Part Two, note 157.

⑰ See Nikos Papadakis, *The International Legal Regime of Artificial Islands* (Sijthoff Publications on Ocean Development, 1977), at 68.

⑱ See A. H. A. Soons, Artificial Islands and Installations in International Law, Law of the Sea Institute, University of Rhode Island, Occasional Paper No. 22, July 1974, 14-15.

⑲ Ibid.

⑳ Michael Redfield, supra. note 158, at 54.

seems to focus on the research location, instead of its nature. He also relies on Milner B. Schaefer,[181] who submitted a paper to the Proceedings of the Second Annual Conference of the Law of the Sea Institute, Rhode Island,[182] in 1967. From Schaefer's paper he quoted a statement made by an official of the State Department of the United States which reads:

> I might extend that by saying that research which involves physical contact with or into the shelf is shelf research and that which does not touch the shelf is not. Thus, measurements of magnetic fields of gravity, or the taking of acoustic subbottom reflection measurements, or water samples would not be considered to be shelf research.[183]

This might not be the authoritative definition. However, it at least demonstrates a stance on the issue. Redfield cited this statement to support his view that physical contact is the decisive factor for the consent requirement of continental shelf research. He went even further to stress the "double criteria currently in use", which means the formulation of continental shelf research requiring "physical contact with the shelf and action on or study of the shelf", was one that requires more than physical contact alone.[184] He also quoted a telegram from the Soviet Union on August 6th, 1969, and a definition of shelf research that it adopted. Both of them justified his view mentioned above.[185] In the telegram, continental shelf

---

[181] He is the director of Institute of Marine Resources, University of California at San Diego.

[182] The Second Conference of the Law of the Sea Institute was held in the University of Rhode Island, on June 26, 1967. The topic is "The Freedom of the Sea's Resources". A chief goal of the conference is to bring together people concerned with the Law of the Sea and to discuss freely in a non-political environment.

[183] See Michael Redfield, supra. note 158, at 54.

[184] Ibid., at 55.

[185] Ibid. The main content of the telegram was to refuse the permission for the R/V Thomas G. Thompson which was owned by University of Washington to conduct research on the Soviet continental shelf.

research was addressed:

> Continental [S]helf research considered to be taking actual samples from the shelf, as by coring, dredging and any other activity which includes physical contact with and action on the shelf. Research on water above the shelf not affected, nor is research with electrical instruments precluded if no physical contact with shelf.[186][sic]

The Soviet Union defined the shelf research as follows:

> Work relating to research, exploration and exploitation of the natural resources of the continental shelf of the USSR, if it must be conducted on the surface or in the subsoil of the seabed of submerged areas [...] shall be subject to registration.[187]

Consequently, both documents emphasized not only "physical contact" with the shelf, but also "actions on" the shelf. Viewed from this perspective, the rather narrow definition of shelf research adopted by the US and USSR, to a certain extent, limited the scope of research subject to the consent requirement.

It is therefore proposed that the second method of interpretation[188] mentioned earlier covered a broad range of shelf research that is subject to the consent requirement and might create numerous inconveniences to the researching State. However, it safeguards the coastal States' interests by preventing the continental shelf from being explored under the guise of scientific research. From today's (post-LOS Convention) perspective, it is

---

[186] See Michael Redfield, supra. note 158, at 55.

[187] Ibid. Here author made a note in Notes and References, note 62. It said that this definition was from "USSR Decree on the Procedure of Conducting Work on the Continental Shelf and the Protection of its Natural Resources of July 18, 1969, (1) [translation by William E. Butler appearing in Soviet Statutes and Decisions 975 (1970)]".

[188] This indicates the interpretation which means the research concerning the Continental Shelf—regardless of it being carried out on the seabed or in the superjacent water—and the research which is conducted on the seabed, whether or not it concerns the Continental Shelf.

difficult to identify clear-cut criteria as to what interpretation of the CCS is right or wrong. Indeed, the ambiguities involved the issue lead to far-reaching amendments to the CCS regime that are today codified in the LOS Convention.

C. The 1982 United Nations Convention on the Law of the Sea

In the LOS Convention, MSR activities are subordinate to a specific legal regime contained in Part XIII. During the negotiations, there were much more demands for control over MSR compared to the former Geneva Conventions.[189] Most of these demands came from developing countries. The scientific community, mostly from developed States, took the opposite position, only because they feared that such control would negatively influence the development of marine science.[190] Ultimately, the LOS Convention met the major demands of the developing countries and introduced a standard of control that was more precise and therefore regulated MSR activities in a narrower way.[191] There are two legal bases of State influence over MSR that are, to some extent, distinct in depth and sphere: sovereignty and jurisdiction.[192] Consequently, for a potential MSR

---

[189] See R. R. Churchill and A. V. Lowe, supra. note 91, at 403.
[190] Ibid.
[191] Ibid. at 403-404.
[192] See Montserrat Gorina-Ysern, An International Regime for Marine Scientific Research (Transnational Publishers, Inc., 2003), at 302. It said that "coastal state exercises sovereignty, sovereign rights, and/or jurisdiction over foreign MSR activities". However, "sovereign rights" is mentioned in LOS Convention that the coastal state exercises sovereign rights for the purpose of exploring or exploiting the natural resources within the Exclusive Economic Zone and Continental Shelf. In the Exclusive Economic Zone, sovereign rights are also exercised for the purpose of conserving and managing the natural resources. There is no relevance to the marine scientific research.

project operator, it is vitally important to determine the legal status of the marine areas where the planned project is to be conducted.

1. Sovereignty—Territorial Sea

LOS Convention Article 2 provides that "the sovereignty of a coastal State extends, beyond its land territory and internal waters [...] to an adjacent belt of sea, described as the territorial sea". During the third United Nations Conference on the Law of the Sea (UNCLOS III), the conduct of MSR in the internal waters was never discussed specifically.[133] Even during the preparatory work, no reference to the regime of internal waters was made.[134] As far as MSR is concerned, it was considered to be the same as the regime for the territorial sea. While there is no definition about the word "sovereignty" in LOS Convention Article 2, with respect to the MSR, the notion of "sovereignty" is substantiated by LOS Convention Article 245, which stipulates that:

> **Coastal States, in the exercise of their sovereignty, have the exclusive right to regulate, authorize and conduct marine scientific research in their territorial sea. Marine scientific research therein shall be conducted only with the express consent of and under the conditions set forth by the coastal State.**

Clearly, it can be seen that in the territorial sea, the conduct of MSR is subject to the sovereignty of the coastal States. However, sovereignty in the territorial sea is subject to the regime of innocent passage.[135] LOS Convention Article 19 (2) (j) provides that carrying out "research or survey activities"

---

[133] See Montserrat Gorina-Ysern, An International Regime for Marine Scientific Research (Transnational Publishers, Inc., 2003), at 304.

[134] See A. H. A. Soons, supra. note 116, at 141. Also see Montserrat Gorina-Ysern, Ibid., at 304.

[135] With respect to the regime of innocent passage in the territorial sea, see LOS Convention Part II section 3.

makes the passage non-innocent. As a result, if a ship passes through the territorial sea and carries out research or survey activities, the passage cannot be considered as innocent. Pursuant to LOS Convention Article 25 (1), the coastal State may take the necessary steps to prevent passage that is not innocent in the territorial sea. Moreover, with respect to non-compliance by warships with the laws and regulations of the coastal State, according to LOS Convention Article 30, if any warship does not comply with coastal State's laws and regulations concerning passage through the territorial sea and disregards any request for the compliance therewith which is made to it, the coastal State may require it to leave the territorial sea immediately. The problems relevant to the phrase "research or survey activities" ought to be assessed since conducting hydrographic surveys is one of the most common "applied research" categories and could potentially fall within the legal limits set by the innocent passage regime.

The question of legitimacy of "survey activities" conducted during innocent passage has a long history of discussion. Under the Geneva Convention on the Territorial Sea and the Contiguous Zone 1958 (CTC 1958),[19] interpretations relating to "passage" and "innocent" were given. CTC 1958 Article 14 (2) provided that "[p]assage means navigation through the territorial sea for the purpose either of traversing that sea without entering internal waters, or of proceeding to internal waters, or of making for the high seas from internal waters". In normal circumstances, the passage shall be continuous and expeditious. According to CTC 1958 Article 14 (3), passage includes stopping and anchoring only in so far as they are incidental to ordinary navigation or are rendered necessary by force majeure or distress. CTC 1958 Article 14 (4) provided that "[p]assage is innocent so long as it is not prejudicial to the peace, good order or security of the coastal State. Such passage shall take place in conformity with these articles and

---

[19] See http://untreaty.un.org/ilc/texts/instruments/english/conventions/8_1_1958_territorial_sea.pdf. (last visited on 14 April 2013).

with other rules of international law". Also, CTC 1958 Article 14 (5) stated that "[p]assage of foreign fishing vessels shall not be considered innocent if they do not observe such laws and regulations as the coastal State may make and publish in order to prevent these vessels from fishing in the territorial sea". It can be seen that in light of CTC 1958 Article 14 (5), survey activities were not even mentioned in the innocent passage context. Thus, the only criterion that could be used to evaluate whether survey activities rendered a passage non-innocent was CTC 1958 Article 14 (4). Perhaps not surprisingly, different interpretative opinions had been expressed in the academic research context. On one hand, some writers held the view that "so long as it is not prejudicial to the peace, good order or security of the coastal State", in conformity with "this convention and with other rules of international law", as well as the laws and regulations adopted by the coastal State relating to innocent passage, the kind of survey activities that were carried out with the ship underway at cruising speed and not involving stopping or anchoring were permissible. [197] It could thus be considered as an exercise of the right of innocent passage[198] with no permission required. [199] On the other hand, according to Redfield, "[o]ther writers while recognizing that some observations and measurements are commonly made without permission or interference from coastal [S]tates, refrain from asserting that

---

[197] See Roger Revelle, Scientific Research on the Sea-Bed: International Cooperation in Scientific Research and Exploration of the Sea-Bed, in: *Symposium on the International Regime of the Sea-Bed* (Instituto Affari Internazionali, 1969), 649-663, at 660. And also see Eberhard Menzel, Scientific Research on the Sea-Bed and its Regime, in: *Symposium on the International Regime of the Sea-Bed* (Instituto Affari Internazionali, 1969), 619-647, at 622.

[198] See Roger Revelle, ibid. He also stated that "this protection probably would not hold if the scientific ship cruised on a grid of survey lines, rather than simply traversing a territorial sea en route to or from the internal waters of the coastal state or to and from the high seas". This made reference to the relationship between hydrographic surveying and the marine scientific research which will be discussed later.

[199] See Eberhard Menzel, supra. note 197.

this may be done as a matter of right"[200]. Meanwhile, one should bear in mind that at that time, even if the survey activities complied with the innocent passage regime, they could still be restricted by the coastal State through enacting national laws and regulations.[201] By virtue of CTC 1958 Article 17, foreign ships exercising the right of innocent passage through the territorial sea were obliged to comply with the laws and regulations adopted by the coastal States. In a 1969 draft revision of the ICO Pamphlet US Oceanic Research in the High Seas and Foreign Waters, the following passage appeared:

> **The United States holds that research vessels need not obtain clearance to operate continuous recording instruments making underway scientific measurements during passage which meets the definition of "innocent passage", through foreign territorial seas unless the coastal State has imposed specific requirements in this regard.**[202]

Thus, coastal States may lawfully influence the conduct concerned. An extreme example of such legislation is the Bulgarian Decree of October 10th, 1951, under which "[f]oreign ships may not engage while in the territorial or internal waters or ports of the People's Republic, in sounding, research, study, photography [...] or make use of radio transmitters, radar, echo sounding or like devices other than those intended for purpose of navigation"[203]. Besides, under CTC 1958 Article 16 (3), coastal States have

---

[200] See Michael Redfield, supra, note 158, at 45 (footnote omitted). Other writers indicate E. D. Brown and W. T. Burke, see E. D. Brown, Freedom of Scientific Research and the Legal Regime of Hydrospace, in: *Indian Journal of International Law*, Vol. 9, 1969, 327-380, at 339. And also see W. T. Burke, *A Report on International Legal Problems of Scientific Research in Oceans* (prepared for the National Council on Marine Resources and Engineering Development, 1967), at 21-24. Excepted from Michael Redfield, supra, note 158, footnote 25, at 90.

[201] Ibid.

[202] Ibid.

[203] See E. D. Brown, supra. note 200, footnote 32, at 338.

the right to temporarily suspend the innocent passage of foreign ships in specified areas of its territorial sea for the sake of security.

Nowadays, with the progress made on the technology and the equipment available, it is possible to conduct more kinds of survey activities, like gravity measurements, investigating the properties of the sea floor and sediments, or making bottom profiles, while the ship is underway at cruising speed without stopping or anchoring. Furthermore, it is similarly possible to make meteorological observations and measurements while underway.[204] As a result, attention has been called to the increasing restrictions placed on MSR and survey activities. Pursuant to the LOS Convention, compared with the CTC 1958, not too much has changed relating to the interpretation of "passage" and "innocent". However, the list of circumstances that can render a "passage" non-innocent has been largely expanded [LOS Convention Article 19 (2)], and survey activities are also expressly listed there.[205] Therefore, from interpretating the LOS Convention, survey activities are not compatible with the innocent passage anymore. Soons observed that:

> If the coastal State has control over access to the territorial sea for the purpose of conducting marine scientific research, it is equally entitled to prohibit any such research to be undertaken during passage. This applies also to hydrographical surveying activities. Only those measurements which are necessary for safe navigation are permitted.[206]

Since the entry into force of the LOS Convention, for a foreign ship who is flying the flag of a State party to the LOS Convention, survey in the coastal States' territorial sea are not allowed anymore and the conducts are supposed to be subject to the coastal States' sovereignty. Likewise,

---

[204] See Eberhard Menzel, supra. note 197, at 622.

[205] LOS Convention Article 19 (2) (j) makes it clear that carrying out of research or survey activities can make the passage non-innocent.

[206] See A. H. A. Soons, supra. note 116, at 47.

conducting hydrographic surveys in the territorial sea is also subject to the sovereignty of coastal States.

Within the regime of innocent passage in the territorial sea, there is another wording problem. LOS Convention Article 21(1) (g) employs the phrases "marine scientific research and hydrographic surveys"; when compared with LOS Convention Article 19 (2) (j), it turns out to have a more restrictive nature.[207] Article 21 stipulates that the coastal State is entitled to adopt laws and regulations relating to innocent passage in the territorial sea. One of the aspects that is regulated by Article 21 (1) (g) is "marine scientific research and hydrographic surveys". In contrast, LOS Convention Article 19, which addresses "meaning of innocent passage", uses "research or survey activities" to stipulate one of the circumstances prejudicial to the peace, good order or security of the coastal State. It literally means that in the territorial sea, domestic laws and regulations relating to innocent passage adopted by the coastal State could only cover a part of the research and survey activities that are comprised as MSR and hydrographic surveys.

Equally, the term "marine scientific research" is also used in LOS Convention Part XIII.[208] In other subject matters, it is referred to in broader terms as "scientific research".[209] LOS Convention Articles 19[210] and 40[211] even refer to any "research". Consequently, "research", "scientific research" and

---

[207] See Montserrat Gorina-Ysern, supra. note 192, at 307.

[208] LOS Convention Part XIII provides legal regime regarding marine scientific research.

[209] Like LOS Convention Article 87,123 and 258-262.

[210] LOS Convention Article 19 stipulates meaning of innocent passage. Article 19 (2) (j) provides that carrying out of research or survey activities shall be considered to be prejudicial to the peace, good order or security of the coastal State.

[211] LOS Convention Article 40 provides that during transit passage in straits used for international navigation, foreign ships, including marine scientific research and hydrographic surveys ships, may not carry out any research or survey activities without the prior authorization of the States bordering straits.

"marine scientific research" are used differently.[212] "Marine scientific research" is the most specific concept and to be regarded as part of "scientific research".[213] The difference between "scientific research" and "marine scientific research" was pointed at by Soons, who noted that:

> Scientific research commonly being regarded as an investigation of a question, problem or phenomenon conducted according to the rules and principles of science, marine scientific research may be regarded as such investigation concerned with the (natural phenomena of the) marine environment.[214]

It can be seen that marine environment becomes a decisive factor here, and that this criterion is used to determine whether a research activity is MSR or scientific research.

The most general word, "research", is used in the context of the regime of innocent passage in the territorial sea, the regime of transit passage of straits used for international navigation,[215] and the regime of archipelagic sea lanes passage.[216] This means that in the territorial sea, in straits used for international navigation, and in the archipelagic waters, ships from other States are subjected to the most intense restrictions. Any research, whether relevant to the marine environment or not, is prohibited during innocent passage, transit passage, and archipelagic sea lane passage. The most specific form of "marine scientific research" is used in LOS Convention Part XIII that covers the territorial sea, the EEZ, and the continental shelf, except section 4, which refers to installations or equipment of all kinds of

---

[212] See A. H. A Soons, supra. note 116, at 124.

[213] See Montserrat Gorina-Ysern, supra. note 192, at 78.

[214] See A. H. A. Soons, supra. note 116, at 124.

[215] In Article 40, even it mentions "marine scientific research and hydrographic survey ships", later it still emphasizes that "any research or survey activities" cannot be carried out without the bordering States consent. Thus, it could be regarded that here it meant to point the general term "research or survey activities".

[216] As regard to the regime of research and survey activities, Article 40 applies to the archipelagic sea lanes passage.

scientific research programs in the marine environment. This implies that the legal regime of LOS Convention Part XIII (except section 4) in principle does not apply to scientific research activities that do not concern the marine environment.[217]

2. Jurisdiction—MSR in the EEZ and on the Continental Shelf

LOS Convention Article 246 (1) expressly prescribes that "[c]oastal States, in the exercise of their jurisdiction, have the right to regulate, authorize and conduct marine scientific research in their EEZ and on their continental shelf in accordance with the relevant provisions of this Convention"[218]. Compared with "sovereignty", which is a central feature of Statehood, "jurisdiction" is regarded as the exercise of State power.[219] The definition of jurisdiction is considered by Florian H. Th. Wegelein to be "the power or competence to legislate and regulate within a certain (geographic or substantive) area and exercise authority over all persons and things within it"[220]. Jurisdiction has generally been treated as a single concept in the international law.[221] However, in recent years three categories have been used: jurisdiction to prescribe (i. e., legislate), jurisdiction to enforce,[222] and jurisdiction to adjudicate.

---

[217] See A. H. A Soons, supra. note 116, at 125.

[218] Article 56.1 (b) also expressly states that states exercise jurisdiction over marine scientific research in exclusive economic zone.

[219] See Florian H. Th. Wegelein, *Marine Scientific Research: The Operation and Status of Research Vessels and other Platforms in International Law* (Martinus Nijhoff Publishers, 2005), at 103 and 105.

[220] Ibid., at 105.

[221] See Oscar Schachter, *International Law in Theory and Practice* (Martinus Nijhoff Publishers, 1991), at 253-254.

[222] Ibid. The same view also expressed in Florian H. Th. Wegelein's book, supra. note 219, at 108. It is said that "in general, jurisdiction entails the authority for legislative (jurisfaction) and executive (adjudication) acts of state; it also includes the competence for judicial review (adjudication)".

(a) Prescriptive Jurisdiction

Legislative jurisdiction is regarded as the State's competence to assert the rights conferred to it by international law[23] and to implement an international treaty or a customary law rule within its domestic legal system.[24] LOS Convention Article 246 provides a consent regime for MSR in the EEZ and on the continental shelf. Article 246 (3) prescribes an obligation for the coastal State to "establish rules and procedures ensuring that such consent will not be delayed or denied unreasonably". It means that in order to implement the consent regime, the coastal State needs to arrange a set of procedures in its domestic law under which the researching State can obtain consent.[25] More importantly, such procedures should prevent applications for consent from being unreasonably delayed or denied. To some extent, such national legislation could effectively prevent the consent regime from being void.[26] Most States have their own domestic laws and regulations over MSR.

(b) Enforcement Jurisdiction

In addition to the jurisdiction to prescribe, the coastal State also has the jurisdiction to enforce its legislation. Generally, enforcement jurisdiction refers to all acts designed to enforce legislative jurisdiction.[27] Enforcement measures are presented in various forms, and these measures are, to some extent, distinct in severity. However, a State successfully enforcing its legislation within its territory or jurisdictional zones does not necessarily mean that the enforcement is legal.[28] The slightest effect can be achieved by

---

[23] See Florian H. Th. Wegelein, supra. note 219, at 108.
[24] Ibid.
[25] Ibid., at 109.
[26] Ibid., at 112.
[27] See Haijiang Yang, *Jurisdiction of the Coastal State over Foreign Merchant Ships in International Waters and the Territorial Sea* (Springer, 2005), at 36.
[28] Ibid., at 37.

measures such as monitoring[229] or approaching.[230] Stopping, boarding, and searching are measures that require physical interference with the vessel's platform.[231] Moreover, the vessel can also be stopped in order to prevent it from entering a certain area or in order to conduct further inspection.[232] The search or inspection of a vessel can easily be linked with intrusive conduct,[233] even when legitimate grounds exist. The severest means include detention,[234] arrest,[235] or seizure.[236] Montserrat Gorina-Ysern expressed that "[i]n some cases, coastal States' domestic implementation of international law rules applicable to all foreign vessels can be considered draconian, particularly if they would be fully enforced"[237]. The lawfulness of enforcement rests not only on the presence of legislative jurisdiction, but also on the propriety of the measures taken and the procedures undergone for the purpose of enforcement.[238] The specific enforcement action should be proportionate to the

---

[229] See Florian H Th. Wegelein, supra. note 219, at 113. It is said that "[t]he objective of Monitoring is to gather evidence of an illegal activity or merely to accumulate information and eventually intelligence about such an activity".

[230] Ibid.. It is said that "[a]pproaching, denoting that the authorities advance the platform for purposes of establishing identify and nationality and a detailed visual scrutiny without impeding safety and passage; as a matter of practice radio communications are established to obtain additional information such as last port of call, next port of call, cargo, and other information pertinent to the voyage".

[231] Ibid.

[232] Ibid.

[233] Ibid.

[234] Ibid., at 114. It proviedes that detention "[d]enotes the act of keeping back, restraining or withholding, either accidentally or by design, a person or thing".

[235] Ibid.. It addresses that "[a]rrest means to deprive a person or a chattel of its liberty by legal authority,[...], into custody for the purpose of holding or detaining them or it to answer a criminal or administrative charge (or civil claim)".

[236] Ibid.. It is said that "[a] seizure is the act of taking possession of property, [...]. The term implies a taking or removal of something from the possession, actual or constructive, of another person or persons".

[237] See Montserrat Gorina-Ysern, supra. note 192, at 17.

[238] See Haijiang Yang, supra. note 227, at 38.

extent of necessary and reasonable to control and punish the offences.[239] In the case of employing force, it was addressed by O'Connell that "there must be adequate warning and instruction, which includes internationally recognized visual signals and sound signals".[240] Additionally, according to LOS Convention Article 225,[241] in the exercise of enforcement jurisdiction against foreign vessels under the LOS Convention, States shall not endanger the safety of navigation or otherwise create any hazard to a vessel, or bring it to an unsafe port or anchorage, or expose the marine environment to an unreasonable risk. Needless to say that some other international law general rules, such as the prohibition of discrimination and the abuse of rights, should be observed as well.[242] Last but not least, enforcement measures may only be exercised by officials or by warships, military aircraft, or other ships or aircraft on government service that shall be identifiably indicated as such.[243]

As far as MSR is concerned, contrary to the enforcement provisions and the safeguards found in the LOS Convention Part XII,[244] LOS Convention Part XIII contains no specific provisions relating to the enforcement of coastal States' regulations and authorisations.[245] Consequently, the Convention's general provisions regarding coastal States' enforcement rights are applicable

---

[239] See Haijiang Yang, supra. note 227, at 38-39.

[240] See D. P. O'Connell, *The International Law of the Sea*, Vol. II (Clarendon Press, 1982), at 1072.

[241] LOS Convention Article 225 stipulates the duty to avoid adverse consequences in the exercise of the powers of enforcement.

[242] See Haijiang Yang, supra. note 227, at 39.

[243] Ibid.

[244] LOS Convention Part XII provides legal regime regarding protection and preservation of the marine environment.

[245] See Myron H. Nordquist (ed.), *United Nations Convention on the Law of the Sea 1982: A Commentary*, Vol. IV, supra. note 110, at 436.

to the matters of MSR,[246] such as the right of hot pursuit.[247] Pursuant to LOS Convention Article 246, the coastal State is not only entitled to regulate and conduct its own MSR, but it may also authorize other States to conduct MSR in its EEZ and on the continental shelf. Therefore, at the domestic level, the coastal State's basic requirement for enforcing MSR regulations is to process the request of the researching State according to the domestic laws and regulations and to ensure that the request is not unreasonably delayed or denied. The coastal State needs to prescribe conditions under which consent can be granted. These conditions generally relate to details of the research project, research platforms, and the personnel participating the research project.[248] Additionally, an authority charged with processing the research request should be set up as well.[249] As a result of the coastal State's prescriptive jurisdiction, denial would be the likely consequence when the conditions set forth by coastal State are not fulfilled by the researching State. Where a research request is denied, the coastal State must therefore be able to enforce the decision to make sure that there is no research activity taking place.[250]

Moreover, LOS Convention Articles 253 (1) and (2) provide that the coastal State enjoys the right to require the suspension or cessation of the MSR activities if they are not conducted in accordance with the information communicated, as provided under LOS Convention Article 248 or Article 249 concerning the coastal State's rights relating to the MSR project. Thus, LOS Convention Article 253 could be regarded as a necessary complement to

---

[246] See Myron H. Nordquist (ed.), *United Nations Convention on the Law of the Sea 1982: A Commentary*, Vol. IV, supra. note 110, at 436.

[247] Ibid. 1982 UNCLOS, Article 111 provides the coastal State's right of hot pursuit for violations of its laws and regulations. Paragraph 1 deals with violations in coastal State's internal waters, archipelagic waters, territorial sea, or contiguous zone. Paragraph 2 deals with violations in the exclusive economic zone or on the continental shelf.

[248] See Florian H. Th. Wegelein, supra. note 219, at 109.

[249] Ibid.

[250] Ibid., at 112.

implement Articles 248 and 249. In this sense, LOS Convention Article 253 "mitigates the possible impact of enforcement measures by the coastal State on the research project in question". [51] "Suspension" means that the research project is interrupted but may be continued if certain requirements or conditions are met. [52] The cessation, however, means that the research project is abandoned, [53] a complete termination, which could be seen as a higher degree of sanction. [54]

(c) Judicial Jurisdiction

Normally, enforcement jurisdiction can be divided further into two types, executive jurisdiction and judicial jurisdiction. [55] Judicial jurisdiction means to subject persons or things to the process of the courts or administrative tribunals of a State. [56] However, in respect of MSR, applications of the judicial jurisdiction are scarce. This is because of the existence of LOS Convention Article 264 according to which disputes concerning the interpretation or application of the provisions of LOS Convention with regard to MSR shall be settled in accordance with LOS Convention Part XV, section 2 and section 3. [57] In this sense, coastal State's judicial jurisdiction does not apply to MSR. Rather, MSR dispute is subject to LOS Convention Part XV, section 2 and 3.

Pursuant to LOS Convention Part XV, section 2 and 3, unless limitations and exceptions stipulated under LOS Convention Part XV section

---

[51] See Myron H. Nordquist (ed.), *United Nations Convention on the Law of the Sea 1982: A Commentary*, Vol. IV, supra. note 110, at 571.

[52] See Florian H. Th. Wegelein, supra. note 219, at 187.

[53] Ibid.

[54] See Myron H. Nordquist (ed.), *United Nations Convention on the Law of the Sea 1982: A Commentary*, Vol. IV, supra. note 110, at 578.

[55] See Haijiang Yang, supra. note 227, at 36.

[56] See Oscar Schachter, supra. note 221, at 255.

[57] LOS Convention Part XV provides settlement of disputes. Section 2 stipulates compulsory procedures that entail binding decision. Section 3 contrarily addresses limitations and exceptions to applicability of section 2.

3 apply, compulsory jurisdiction of an international court or tribunal comes into play. The specific limitation with respect to MSR contained in LOS Convention Part XV, section 3 Article 297 (2). LOS Convention Article 297 (2) (a) (i) and (ii) provide that disputes concerning the exercise of the discretionary powers of the coastal State to withhold consent, or its decision to order suspension or cessation of a project for a research project in the EEZ or on the continental shelf are excluded from the scope of the compulsory procedures that provided by LOS Convention Part XV section 2. Besides, LOS Convention Article 297 (2) (b) addresses that disputes arising from an allegation that a coastal State is not exercising its rights under LOS Convention Article 246 and 253[58] on a specific project in a manner compatible with the LOS Convention shall be submitted to conciliation[59] at the request of any party. Following this line of consideration, based on LOS Convention Part XV section 2 and 3, disputes settlement relating to MSR can be further subdivided into three categories: one is the general compulsory procedures which can entail binding decisions. The second one is that, the compulsory procedures can be launched, however, due to the discretionary character, according to Article 297 (2) (a) (i) and (ii), the coastal State shall not be obliged to accept the submission.[60] That is to say, under the circumstances which described by Article 297 (2) (a) (i) and (ii), the compulsory procedures can be void.

The last one is the compulsory conciliation procedure which adopted under the circumstances indicated in LOS Convention Article 297 (2) (b).[61]

---

[58] LOS Convention Article 253 addresses legal regulation regarding suspension or cessation of marine scientific research activities.

[59] Annex E to the LOS Convention provides legal regime with respect to conciliation. Annex E section 2 specifically deals with compulsory submission to conciliation procedure pursuant to LOS Convention Part XV section 3.

[60] See Lous B. Sohn, John E. Noyes, *Cases and Materials on the Law of the Sea* (Transnational Publishers, 2004), at 801.

[61] Ibid.

3. Freedom—MSR in the Areas beyond National Jurisdiction

In addition to the sovereignty and jurisdiction of a State, the LOS Convention recognizes States' freedom to conduct MSR beyond the limits of national jurisdiction.[262] Part XIII contains two provisions dealing specifically with the regime for MSR undertaken in the marine areas beyond the national jurisdiction. The first provision is Article 256, which deals with MSR in the "Area", and the second provision is Article 257 which covers MSR in the water column beyond the EEZ.

(a) MSR in the "Area"

"Area" means the seabed and ocean floor and subsoil thereof, beyond the limits of national jurisdiction.[263] The regime of the Area is laid down in LOS Convention Part XI. In respect of MSR in the Area, LOS Convention Article 143 establishes the general proposition that MSR in the Area shall be carried out exclusively for peaceful purposes and for the benefit of mankind as a whole, in accordance with Part XIII. While it allows both State parties and the Authority to conduct MSR, it also sets out the obligations of States parties and the Authority to encourage international cooperation in MSR and the dissemination of MSR results and analysis, as well as enhancing developing States' capacity-building. Besides that, LOS Convention Part XIII Article 256, grants all States and competent international organizations the right to conduct MSR in the Area with the requirement to act in conformity with the provisions of Part XI. For the sake of a better understanding, three issues thus need to be clarified: first, whether MSR can be considered as an activity in the Area. Second, how to understand MSR in the Area. Third, who enjoys the right to conduct MSR in the Area.

LOS Convention Article 134 (2) makes it clear that activities in the Area shall be governed by the provisions of LOS Convention Part XI.

---

[262] See Montserrat Gorina-Ysern, supra. note 192, at 318.
[263] LOS Convention Article 1(1) (1).

According to LOS Convention Article 1 (3), the expression "activities in the Area" means all activities of exploration for and exploitation of the resources of the Area. Furthermore, based on LOS Convention Article 133,[264] the term "resources" means mineral resources, whether they are solid, liquid or gaseous. They are referred to as "minerals" when they are extracted from the Area. That is to say, whether MSR can be considered as an "activity in the area" depends on whether the research is in direct relation to the exploration for or exploitation of the Area's mineral resources. If it does, then it is subject to LOS Convention Part XI. LOS Convention Article 151 provides that activities in the Area shall be carried out by the Authority,[265] and that those activities are required to be conducted on the Authority's behalf. Therefore, if the MSR is considered to be an "activity in the Area", it becomes subject to the Authority's complete regulatory control.[266] Besides, it is worth noting that even though LOS Convention Article 143 specifically stipulates legal regulations regarding MSR in the Area, there is no way to be certain that Article 143 applies to all kinds of MSR, no matter it is an activity in the Area or not. This is because of the implication of LOS Convention Article 140 which addresses "benefit of mankind" due to the wording of LOS Convention Article 143 (1).[267] Pursuant to LOS Convention

---

[264] LOS Convention Article 133 reads that "for the purpose of this Part: (a) 'resources' means all solid, liquid or gaseous mineral resources in situ in the Area at or beneath the sea-bed, including polymetallic nodules; (b) resources, when recovered from the Area, are referred to as 'minerals'".

[265] According to the LOS Convention Article 156 and 157, the Authority is named as the International Seabed Authority, which is a new organization that organizes and controls activities in the Area, particularly with a view to administering the resources of the Area.

[266] See Wesley S. Scholz, Oceanic Research—International Law and National Legislation, in: *Marine Policy*, Vol. 4, 1980, 91-125, at 113.

[267] LOS Convention Article 143 (1) reads: "Marine scientific research in the Area shall be carried out exclusively for peaceful purposes and for the benefit of mankind as a whole, in accordance with Part XIII." See Alexander Proelss, Marine Genetic Resources under UNCLOS and the CBD, in: *German Yearbook of International Law*, Vol. 51, 417-446, at 424.

Article 140,[268] "benefit of mankind" formula applies to activities in the Area. It thus leads LOS Convention Article 143 to arguably also apply to the MSR that is in direct relation to the exploration for or exploitation of the Area's mineral resources.[269]

At first sight, the interpretation of "marine scientific research in the Area" as stipulated in LOS Convention Article 143 and 256 seems vague in terms of how to interpret "research in the Area". Does it mean the research physically conducted in the Area, or the research concerning the seabed and subsoil conducted in the superjacent waters, without involving physical contact with the seabed and subsoil? LOS Convention Article 257 covers MSR conducted in the waters superjacent to the Area.[270] This means that while MSR physically conducted in the Area is governed by LOS Convention Article 143 and 256, MSR concerning the Area but conducted in superjacent waters is governed by LOS Convention Article 257. Thus, "marine scientific research in the Area" should be interpreted as the research that is physically conducted in the Area.

One question might be raised when LOS Convention Article 143 is read

---

[268] LOS Convention Article 140 is about benefit of mankind. It stipulates that "1. Activities in the Area shall, as specifically provided for in this Part, be carried out for the benefit of mankind as a whole, irrespective of the geographical location of States, whether coastal or land-locked, and taking into particular consideration the interests and needs of developing States and of peoples who have not attained full independence or other self-governing status recognized by the United Nations in accordance with General Assembly resolution 1514 (XV) and other relevant General Assembly resolutions. 2. The Authority shall provide for the equitable sharing of financial and other economic benefits derived from activities in the Area through any appropriate mechanism, on a non-discriminatory basis, in accordance with Article 160, paragraph 2(f)(i)".

[269] This issue is further analysed, see infra. Chapter III, C (3) (c).

[270] LOS Convention Article 257 provides that all States and competent international organizations have the right, in conformity with the Convention, to conduct marine scientific research in the water column beyond the limits of the exclusive economic zone. See Florian H. Th. Wegelein, supra. note 219, at 208. And the same view see A. H. A. Soons, supra. note 116, at 227.

together with LOS Convention Article 256: whether the right to conduct MSR in the Area is enjoyed by all States and competent organizations, as provided in Article 256 of the LOS Convention or only by the Authority and State parties to the Convention, as provided in LOS Convention Article 143. [271] A definite answer cannot be given since the wording is somewhat confusing. [272] Furthermore, Soons proposed that it cannot be simply argued that LOS Convention Article 143 "lays down further requirements limiting the scope of the subjects of the rights to conduct research". [273] However, one thing is certain: LOS Convention Articles 143 and 256, when read together, empower the Authority as a competent international organization with the right to conduct MSR. [274]

LOS Convention Article 143 grants both the Authority and State parties the right to conduct MSR in the Area. In conjunction with LOS Convention Article 256, it is recognised that "states do in fact have an independent right to carry on research activities in the Area". [275] Furthermore, three qualifications should be fulfilled when MSR is carried out in the Area: (i) it should be exclusively for peaceful purposes, (ii) it should be for the benefit of mankind as a whole, and (iii) it should be in accordance with LOS Convention Part XIII. [276] Apart from that, States that carry out MSR need to fulfil the obligation to promote international cooperation in MSR in the Area. This can be achieved through participation in international programmes, through support of developing countries in their endeavours,

---

[271] See A. H. A. Soons, supra. note 116, at 227.
[272] Ibid.
[273] Ibid.
[274] Ibid.
[275] See Wesley S. Scholz, supra. note 266, at 114. The same view see A. H. A. Soons, supra. note 116, at 228.
[276] Wesley S. Scholz expressed that in his article: "[B]y its terms research shall be carried on in accordance with Part XIII, Article 257 of which confirms the existence of a 'right' for states 'to conduct marine scientific research in the Area in conformity with the provisions of Part XI'. The references are circular and result in ambiguity, [...]."

and through disseminating research results.[277] Although the Authority and State parties are both entitled to carry out MSR in the Area, according to LOS Convention Article 143, however, on closer examination a slight difference appears within the formulation of Article 143 (2) and (3). While LOS Convention Article 143 (2) provides that "[t]he Authority may carry out marine scientific research concerning the Area and its resources, [...]", LOS Convention Article 143 (3) states that "States Parties may carry out marine scientific research in the Area. [...]". The phrase "concerning the Area and its resources" within LOS Convention Article 143 (2) means that research is permitted as long as the research concerns the Area and its resources, regardless of whether the research is physically undertaken in the Area or not. As for the Authority, the research can be undertaken freely if the research concerns the Area and its resources, no matter where it is conducted. However, for State parties, the research can only be physically conducted in the Area, regardless of whether it concerns the Area and the resources or not.

(b) MSR in the Water Column beyond the EEZ

Based on LOS Convention Article 86,[278] the water column beyond the EEZ is the high seas. LOS Convention Article 87 (1) (f) affirms that MSR belongs to the freedom of the high seas. Yet, the freedom has to be exercised subject to the requirements of LOS Convention Parts VI and XIII. The reason why LOS Convention Article 87 (1) (f) makes reference to Part VI is because it "has direct significance for the conduct of marine scientific research only where a coastal State, in accordance with LOS Convention Article 76, has successfully claimed an outer continental shelf"[279]. In this

---

[277] See Florian H. Th. Wegelein, supra. note 219, at 209.

[278] It is expressed that the provisions of this part apply to all parts of the sea that are not included in the exclusive economic zone, in the territorial sea or in the internal waters of a State, or in the archipelagic waters of an archipelagic State. This article does not entail any abridgement of the freedoms enjoyed by all States in the exclusive economic zone in accordance with Article 58.

[279] See Florian H. Th. Wegelein, supra. note 219, at 206.

area there could be conflicts between the coastal State's rights regarding the outer continental shelf and the researching State's rights under the high seas regime.[20] LOS Convention Article 76 addresses the possibility that the coastal State could establish an "outer" continental shelf extending beyond 200 nautical miles from the baselines.[21] MSR on such an outer continental shelf is subject to the coastal State's jurisdiction since, according to LOS Convention Article 246 (6), the coastal State has a right to prior consent in those areas that it has designated for exploration and exploitation. However, one should bear in mind that the coastal State's rights over the continental shelf do not affect the legal status of the superjacent waters.[22] The water column above the outer continental shelf still belongs to the high seas. Moreover, according to LOS Convention Article 257,[23] MSR conducted in the water column beyond the EEZ, which has nothing to do with the outer continental shelf, could be conducted and is not confined by LOS Convention Article 246 (6). However, the research still has to be done in conformity with the LOS Convention.

4. Hydrographic Surveys

Another intractable problem related to MSR that the LOS Convention is facing is the term hydrographic surveys. There are four general categories of marine data collection, according to Ashley Roach, which are respectively MSR, surveys, operational oceanography and exploration and exploitation of

---

[20] See Florian H. Th. Wegelein, supra. note 219, at 206.
[21] It will be analysed in the next section.
[22] LOS Convention, Article 78.
[23] Article 257 is about the marine scientific research in the water column beyond the exclusive economic zone. It is said that "[a]ll States, irrespective of their geographical location, and competent international organizations have the right, in conformity with this Convention, to conduct marine scientific research in the water column beyond the limits of the exclusive economic zone".

natural resources and underwater cultural heritage.[284] Surveys are often presented in two forms, hydrographic surveys and military surveys.[285]

A hydrographic survey was defined by the International Hydrographic Organization(IHO)[286] in the International Hydrographic Dictionary as:

[a] survey having for its principal purpose the determination of data relating to bodies of water. A hydrographic survey may consist of the determination of one or several of the following classes of data: depth of water, configuration and natural of the bottom; directions and force of currents; heights and times of tide and water stages; and location of topographic features and fixed objects for survey and navigation purpose.[287]

This definition focuses on the nature and types of marine data collection rather than on the purpose of data collection.[288] The traditional function of a hydrographic survey is to obtain information for creating navigational charts and for navigational safety.[289]

In contrast, military surveys are activities that involve the collection of marine data for military purposes.[290] The data collected may be in classified or unclassified form and normally is not available to the public or the

---

[284] See Ashley Roach, Marine Data Collection: Methods and the Law, in Myron H. Nordquist, Tommy T. B. Koh and John Norton Moore (eds.), *Freedom of Seas, Passage Rights and the* 1982 *Law of the Sea Convention* (Martinus Nijhoff Publishers, Leiden/Boston, 2009), 171-208, at 173.

[285] Ibid.

[286] The International Hydrographic Organization is a technical organization that provides national hydrographic services. It focused on maintaining the highest standards of hydrograph and charting, and ensuring common standards of nautical charting.

[287] Excerpted from Sam Bateman, Hydrographic Surveying and Marine Scientific Research in Exclusive Economic Zones, in: Myron H. Nordquist, Tommy T. B. Koh and John Norton Moore (eds.), *Freedom of Seas, Passage Rights and the* 1982 *Law of the Sea Convention* (Martinus Nijhoff Publishers, Leiden/Boston, 2009), 105-132, at 114.

[288] Ibid.

[289] See Ashley Roach, supra. note 284, at 175.

[290] Ibid. Also see Sam Bateman, supra. note 287, at 114-115.

scientific community unless it is unclassified or was collected on the high seas.[280] Notwithstanding that hydrographic surveys are therefore different from military surveys, survey activities comprise both hydrographic surveys and military surveys.

Before taking a closer look at hydrographic surveys, one thing needs to be reiterated: according to the express stipulation of LOS Convention Article 19 (2) (j), "carrying out research or survey activities" in the territorial sea shall be considered non-innocent. Therefore, conducting hydrographic surveys will be subject to the sovereignty of the coastal States. All further analyses relating to hydrographic survey are considered under the context of conducting surveys in any marine areas other than the territorial sea.

LOS Convention Article 21(1) (g) vests coastal States with the right to adopt laws and regulations with respect to "marine scientific research activities and hydrographic surveys". Since the LOS Convention does not use the term "marine scientific research" to cover "hydrographic surveys", it has led scholars to debate the relationship between MSR and hydrographic surveys. The question is whether conducting hydrographic surveys should be regarded as a part of MSR and thus subject to the regime for MSR established in LOS Convention Part XIII.

Notably however, MSR is identical to hydrographic surveys insofar as they both emanate from marine data collection. Notwithstanding this, Ashley Roach favours the view that hydrographic surveys are not MSR, and that conducting hydrographic surveys in the EEZ, continental shelf, high seas, or the Area is a high seas freedom relevant to the operation of ships and aircrafts.[281] Likewise, Soons considers that in spite of hydrographic surveying being somewhat similar to MSR, they still cannot be regarded as such. Soons contends that:

> **the collection of data by a ship (be it a research vessel or not) in passage that are required for the safe passage of the ship**

---

[280] See Ashley Roach, supra. note 284, at 175.
[281] Ibid., at 180.

in question (such as water depth, wind speed and direction) cannot be considered to be "research". It must be regarded a normal activity associated with the operation of ships. [283]

As long as hydrographic surveys are conducted for the navigational safety or in connection with laying submarine cables or pipelines, rather than for the purpose of exploration or exploitation of natural resources, on the basis of LOS Convention Article 58 (1) it must be regarded as an internationally lawful use of the sea that can be freely conducted in the EEZ, on the continental shelf, on the high seas or in the Area. [284]

Contrarily, Sam Bateman argues that "[n]ot only state practice but also discussion in expert bodies would seem to support the argument that hydrographic surveying in the EEZ should now be regarded as marine scientific research". [285] In addition, he also points out that the IHO "regards hydrographic surveying as part of marine scientific research and thus subject to the regime of marine scientific research established in Part XIII of UNCLOS [LOS Convention]". [286]

Against this background, it seems difficult to draw a conclusion on this issue. Given that the LOS Convention neither contains a definition of MSR nor of hydrographic survey, States are relatively free to find their own interpretations and accordingly employ different systems of implementing control. [287] In the case of the United States, it is addressed that:

> **While the Law of the Sea Convention does not define marine scientific research (MSR), the term generally refers to those activities undertaken in the ocean to expand knowledge of the marine environment and its processes. The United States has identified some marine data collection activities that are not**

---

[283] See A. H. A. Soons, supra. note 116, at 149.
[284] Ibid., at 157.
[285] See Sam Bateman, supra. note 287, at 124.
[286] Ibid., at 125.
[287] See Florian H. Th. Wegelein, supra. note 219, at 81.

marine scientific research. These include prospecting for and exploration of natural resources; hydrographic surveys (for enhancing the safety of navigation)(italic added); ...㉘

Likewise, during the 25th International Hydrographic Conference held in 1997, while France proposed that "the general provisions laid down by the UN Convention on the Law of the Sea [LOS Convention] as regards marine scientific research and exploration of the seabed apply to hydrographic surveys", the UK responded as follows: "The UK cannot support this proposal. It is considered that hydrographic surveys are not, and should not be, governed by UNCLOS [LOS Convention] Part XIII."㉙

In contrast, in 1996, China promulgated a Regulation on the Management of Foreign-Related Marine Scientific Research (hereinafter China Regulations).㉚ Article 2 of the China Regulations stipulates that:

> [t]hese Regulations shall apply to the conduct by international organizations, organizations and individuals of any foreign country of survey activities (italic added) and research on marine environment and marine resources in the internal sea, territorial sea and other sea areas under the jurisdiction of the People's Republic of China solely for peaceful purposes by themselves or in cooperation with Chinese organizations.㉛

It is thus can be seen that under the China Regulations, survey activities are subject to the same legal regime which prescribed for MSR. Furthermore, Australia and Canada are understood to seek permission of the

---

㉘ See Marine Scientific Research Authorizations, available at http://www.state.gov/e/oes/ocns/opa/rvc/index.htm (last visited on 5 August 2013).

㉙ See Geoffrey Marston, United Kingdom Materials on International Law 1997, in: *The British Yearbook of International Law*, Vol. 68, 1997, 467-644, at 609.

㉚ See http://www.soa.gov.cn/zwgk/fwjgwywj/shxzfg/201211/t20121105_5233.html (Chinese version) (last visited on 5 August 2013). English version available on http://wdc-d.coi.gov.cn/fg/wgl2.htm (last visited on 5 August 2013).

㉛ See Article 2 of Regulation on the Management of Foreign-Related Marine Scientific Research.

coastal State before conducting hydrographic surveys in the EEZ of another State.[382] This indicates that the national legislation of these countries does not distinguish between MSR and hydrographic surveys.[383]

Consequently, following this line of consideration, researchers are advised to consult with the relevant authorities even if an official request may not be necessary prima facie since coastal States may employ a different reading or may even claim jurisdiction over hydrographic surveys.[384]

5. 1982 LOS Convention Article 246

All prior discussions dealt with the different levels of State influence on MSR, which may be summarized by reference to the concepts of "sovereignty", "jurisdiction" and "freedom". However, the legal regime's contents of MSR in the EEZ and on the continental shelf, which are both included in LOS Convention Article 246, should be given further consideration. Notably, "Article 246 is the only article in the Convention to address simultaneously the balancing of the substantive rights of the coastal State and other States both in the EEZ and on the continental shelf (the orientation of Articles 111, 210, 216, 248, 249 and 253, paragraph 1, which also refers to both zones, is quite different and the expressions are used in their primary geographical meanings)"[385].

(a) Jurisdiction without "Exclusive"

LOS Convention Article 246, when compared to Article 245, which allows coastal States to exercise their sovereignty and grants them the exclusive right to regulate, authorize and conduct MSR in their territorial sea, also provides the coastal States with the right to regulate, authorize and conduct MSR. However, what is missing is the word "exclusive" in relation

---

[382] See Sam Bateman, supra. note 287, at 124.
[383] Ibid.
[384] See Florian H. Th. Wegelein, supra. note 219, at 81.
[385] See Myron H. Nordquist (ed.), *United Nations Convention on the Law of the Sea 1982: A Commentary*, Vol. IV, supra. note 110, at 517.

to the exercise of their jurisdiction in their EEZ and on the continental shelf. The word "exclusive" makes a significant difference.

First, turning to the denotation of the word "exclusive", LOS Convention Article 77 concerns coastal States' rights over the continental shelf and accords coastal States with sovereign rights for the purpose of exploring the continental shelf and exploiting its natural resources. LOS Convention Article 77 (2) provides that "[t]he rights referred to in paragraph 1 are exclusive in the sense that if the coastal State does not explore the continental shelf or exploit its natural resources, no one may undertake these activities without the express consent of the coastal State". Wegelein points out that "[i]t precludes concurrent jurisdiction or exercise of activities. 'Exclusive' in the sense of Article 77(2) means that non-regulation is regulation, namely, no activity at all"[306].

Second, LOS Convention Articles 246 and 56 (1) (b) also make reference to MSR. LOS Convention Article 56 (1) (b) clarifies coastal States' jurisdiction in the EEZ regarding three aspects: (i) the establishment and use of artificial islands, installations and structures; (ii) MSR; and (iii) the protection and preservation of the marine environment. However, LOS Convention Article 60 (b)[307] makes a further emphasis by using the phrase "exclusive jurisdiction" only with respect to artificial islands, installations and structures, but does not refer to MSR and the protection and preservation of the marine environment,[308] which are dealt with in Part XIII

---

[306] See Florian H. Th. Wegelein, supra. note 219, at 185.

[307] The first paragraph of Article 60 provides the exclusive rights of the coastal State in the Exclusive Economic Zone to construct and to authorize and regulate the construction, operation and use of artificial islands; installations and structures for the purposes provided in the Article 56 and other economic purposes; installations and structures which may interfere with the exercise of the rights of the coastal State in the zone. Article 60 (b) provides that "[t]he coastal State shall have exclusive jurisdiction over such artificial islands, installations, and structures, including jurisdiction with regard to customs, fiscal, health, safety, and immigration laws and regulations".

[308] See Florian H. Th. Wegelein, supra. note 219, at 184.

and Part XII, respectively. Wegelein argues that "Article 56 (1) (b) (ii) gives the coastal State only a prerogative or pre-emptive right. Where the coastal State does not avail itself of the possibilities offered by the 1982 LOS Convention, other States may operate on the assumption of the high seas freedoms"[309]. This is deduced from the comparison with the case of artificial islands, installations and structures over which the coastal States require "exclusive jurisdiction".

As mentioned above, LOS Convention Article 246 in Part XIII does not use the phrase "exclusive jurisdiction" in order to strike a balance between coastal States' extended control and the traditional freedoms of the high seas.[310] It could be "viewed as a concession to the previous freedom of marine scientific research in the waters now under coastal State jurisdiction"[311].

Finally, it should be noted that in contrast to LOS Convention Part V,[312] Part VI, which deals with the regime of the continental shelf, makes no reference to MSR on the continental shelf. This seems inconsistent, especially since LOS Convention Article 56 (1) (b) expressly mentions jurisdiction over MSR.[313]

(b) The Consent Regime

LOS Convention Article 246 (2) provides that "[m]arine scientific research in the exclusive economic zone and on the continental shelf shall be conducted with the consent of the coastal State". The regime is clearly elaborated by Nordquist, who notes that:

> **[t]he compromise embodied in article 246 consists of acceptance of the principle of coastal State consent as the norm, coupled with the specific exceptions contained in paragraph 5.**

---

[309] See Florian H. Th. Wegelein, supra. note 219, at 185.
[310] Ibid., at 184.
[311] See Florian H. Th. Wegelein, supra. note 219, at 185.
[312] Part V is dealing with the legal regime with respect to the Exclusive Economic Zone.
[313] See A. H. A. Soons, supra. note 116, at 159.

II. Marine Scientific Research under the International Law of the Sea

This leads to the concept of qualified consent and the dropping of any attempt to base the regime of a purported differentiation between pure scientific research and scientific research of direct significance for the exploration and exploitation of natural resources in the exclusive economic zone or on the continental shelf.[14]

There was a long, struggled negotiation during the discussions of several Conference sessions regarding the legal principles that should govern the conduct of MSR in the area beyond the territorial sea under coastal State jurisdiction.[15]

During the second session of UNCLOS III in 1974, which dealt, inter alia, with the question whether the conduct of the research should be free or subject to the consent requirement, four trends evolved: the absolute consent regime,[16] the qualified consent regime,[17] the notification regime,[18] and the partial consent regime.[19] No unanimous solution has been reached for this basic issue.

During the third session of UNCLOS III in 1975, an important

---

[14] See Myron H. Nordquist (ed.), *United Nations Convention on the Law of the Sea* 1982: *A Commentary*, Vol. IV, supra. note 110, at 518.

[15] See A. H. A. Soons, supra. note 116, at 159.

[16] The consent of the coastal State would be required for the conduct of marine scientific research in areas under its jurisdiction; the coastal State would have the discretionary power to prohibit such research. See A. H. A. Soons, supra. note 116, at 160.

[17] Coastal State consent would have to be obtained, but this consent should not normally be withheld by the coastal State if certain internationally agreed conditions are met. See A. H. A. Soons, ibid.

[18] Consent of the coastal State would not be required, but those conducting the research would have to notify the coastal State and comply with certain internationally agreed conditions. See A. H. A. Soons, ibid.

[19] Marine scientific research concerned with the exploration or exploitation of natural resources would be subject to an absolute consent regime; all other research would be free. See A. H. A. Soons, ibid.

development was made that could provide the basis for a compromise solution.[120] The development was that based upon distinguishing two categories of the MSR, which are research of a fundamental nature and research related to the resources of the EEZ and continental shelf, different legal regimes should be applied.[121] In the Informal Single Negotiating Text Part Ⅲ (ISNT Part Ⅲ),[122] which was prepared at the request of the Conference, Part Ⅱ MSR (ISNT Part Ⅲ Part Ⅱ MSR) Article 21 provided that when the research project relates to living and non-living resources of the EEZ and the continental shelf, it shall be conducted only under the coastal State's explicit consent. In contrast, ISNT Part Ⅲ Part Ⅱ MSR Article 22 only stated that if the research is of a fundamental nature, the coastal State may indicate its intent to participate in the different phases of the research concerned on mutually agreed terms. In case the coastal State does not reply, the researching State or the international organization shall proceed with meeting all the requirements referred to in ISNT Part Ⅲ Part Ⅱ MSR Article 16.[123] To summarize, if the research is of a fundamental

---

[120] See A. H. A. Soons, supra. note 116, at 160-161.

[121] Ibid.

[122] See http://untreaty.un.org/cod/diplomaticconferences/lawofthesea-1982/docs/vol_IV/a_conf_62_wp-8_part-3.pdf (last visited on 15 April 2013).

[123] Informal Single Negotiating Text (ISNT) Part Ⅲ, Part Ⅱ MSR Article 16: States and international organizations when undertaking scientific research shall comply with the following conditions: (a) Ensure the rights of the coastal State, if it so desired, to participate or to be represented in the research project; (b) Provide the coastal State an opportunity to participate directly or be represented, if it so desires, in the research on board vessels at the expense of the State conducting the research but without payment of any remuneration to the scientist of the coastal State; (c) Provide the coastal State with the final results and conclusions of the research project; (d) Undertake to provide to the coastal State, on an agreed basis, raw and processed data and samples of materials; (e) If requested, assist the coastal State in assessing the said data and samples and the results thereof; (f) Ensure that the research results are made internationally available through International Data Centres or through other appropriate international channels as soon as feasible; (g) Inform the coastal State immediately of any major change in the research programme; and (h) Comply with all relevant provisions of this Convention.

nature, it is subject to the notification regime.[324] If the research is related to the resources of the EEZ and the continental shelf, it is subject to an absolute consent regime.[325]

However, at the fourth session of UNCLOS Ⅲ in 1976, a full consent regime superseded the MSR provisions of the ISNT Part Ⅲ.[326] It denoted that the consent regime applies to all categories of MSR, but if the research related to natural resources, it would be subject to an absolute consent regime.[327] Other MSR would be subject to a qualified consent regime.[328]

In the fifth session of UNCLOS Ⅲ in 1976 relating to this issue, only some wording modifications were done without any further essential improvement. At the end of the fifth session, the Chairman drafted a text that Soons refers to as the "test proposal".[329] In this proposal, the coastal State's consent would be required for all MSR in the EEZ and on the continental shelf. The coastal State should normally grant its consent. It could only withhold its consent if the research (ⅰ) bears upon the exploration and exploitation of the natural resources; (ⅱ) involves drilling, the use of explosives or the introduction of harmful substances into the marine

---

[324] See supra. note 318.

[325] See supra. note 316. See A. H. A. Soons, supra. note 116, at 161.

[326] At the fourth session, a new document was produced called "Revised Single Negotiation Text Part Ⅲ" (RSNT Part Ⅲ). See http://untreaty.un.org/cod/diplomatic-conferences/lawofthesea-1982/docs/vol_V/a_conf-62_wp-8_rev-1-part3.pdf. (last visited on 15 April 2013).

[327] According to Article 60 of RSNT, in addition to the case of research relating to the natural resources, some other circumstances also could lead to the application of the absolute consent regime, namely if the marine scientific research: (a) involves drilling or the use of explosives; (b) unduly interferes with economic activities of the coastal State in accordance with its jurisdiction; or (c) involves the use of artificial islands, installations or structures.

[328] The meaning of qualified consent regime is elaborated by A. H. A. Soons, see supra. note 317. More details see Revised Single Negotiation Text (RSNT) Article 60. See A. H. A. Soons, supra. note 116, at 161.

[329] See A. H. A. Soons, supra. note 116, at 162.

environment; or (iii) involves the use of artificial islands, installations or structures. [30]

In the sixth session of UNCLOS III, the Informal Composite Negotiating Text (ICNT) [31] was produced. The formulation of the consent regime in Article 247 ICNT was almost the same as the "test proposal". [32] Due to the "package deal" at the seventh session, the ICNT text was regarded as acceptable, even though it was met with some reluctance by most delegations. [33] During the ninth session, in addition to some minor drafting changes, three more paragraphs were added in Article 246 of the Informal Composite Negotiating Text Revision 2 (ICNT/Rev. 2). [34] The first paragraph relates to the normal circumstances that might exist despite the absence of diplomatic relations between the coastal and the researching State. [35] The other two paragraphs deal with research on the continental shelf beyond 200 nautical miles. [36] Therefore, Article 246 ICNT/Rev. 2 subsequently became LOS Convention Article 246. [37]

(c) Normal Circumstances

UNCLOS Article 246 (3) provides that:

Coastal States shall, in normal circumstances, grant their consent for marine scientific research projects by other States or competent international organizations in their exclusive economic zone or on the continental shelf to be carried out in accordance with this Convention exclusively for peaceful purposes and in

---

[30] See A. H. A. Soons, supra. note 116, at 162.

[31] See http://untreaty. un. org/cod/diplomaticconferences/lawofthesea-1982/docs/vol_VIII/a_conf-62_wp-10. pdf. (last visited on 15 April 2013).

[32] See A. H. A. Soons, supra. note 116, at 163.

[33] Ibid.

[34] The document is Informal Composite Negotiating Text Revision 2 (ICNT/Rev. 2) dsee A/CONF. 62/WP. 10/Rev. 2. See A. H. A. Soons, supra. note 116, at 163.

[35] The United States of America proposed this paragraph. See ICNT/Rev. 2.

[36] See A. H. A. Soons, supra. note 116, at 163.

[37] Ibid.

order to increase scientific knowledge of the marine environment for the benefit of all mankind. To this end, coastal States shall establish rules and procedures ensuring that such consent will not be delayed or denied unreasonably.

In other words, not all cases could lead to the coastal State granting its consent. The coastal State's duty is not absolute. The duty only applies "in normal circumstances".[338] The expression "in normal circumstances" is the substitute of "normally", which was provided in the "test proposal".[339] At that time, the provision with the word "normally" evoked strong objections on the ground that it was too broad[340] and could easily lead to subjective decisions.[341] On the contrary, the proponents of the phrase "in normal circumstances" argued that "[i]t meets the asserted needs of the coastal State by providing for abnormal diplomatic circumstances between the [S]tates concerned relevant to the project"[342] and introduces an objective criterion in determining the existence of abnormal circumstances.[343] However, Soons points out that "[t]he coastal State always retains a residual power to withhold consent in view of the existence of abnormal, or non-normal circumstances"[344].

Besides deferring to the coastal State's interpretation of "normal circumstances", it should be noted that there are three inflexible restrictions within LOS Convention Article 246 (3) regarding the conduct of MSR.

---

[338] See A. H. A. Soons, supra. note 116, at 167.

[339] Ibid., at 286.

[340] See Bernard H. Oxman, The Third United Nations Conference on the Law of the Sea: The 1977 New York Session, in: *American Journal of International Law*, Vol. 72, 1978, 57É83, at 76-77.

[341] See A. H. A. Soons, supra. note 116, at 168.

[342] See Bernard H. Oxman, supra. note 340, at 77.

[343] See A. H. A. Soons, supra. note 116, at 168.

[344] Ibid., at 167. The same view also shared by Myron H. Nordquist (ed.), *United Nations Convention on the Law of the Sea* 1982: *A Commentary*, Vol. Ⅳ, supra. note 110, at 518.

These are "in accordance with the Convention", "exclusively for peaceful purposes", and "in order to increase scientific knowledge of the marine environment for the benefit of all mankind".⁽³⁶⁾

(i) In Accordance with the Convention

This restriction means conducting MSR should comply with all provisions of the LOS Convention. Generally, the way how coastal States decide whether the conduct of MSR is carried out in accordance with the LOS Convention would be on the basis of a communication with the researching States or international organizations that contain the information required pursuant to LOS Convention Article 248.⁽³⁶⁾ Soons elaborated that even if the provisions of the LOS Convention have not been complied with, "[t]he coastal State cannot exercise a discretion as in the cases specified in paragraph 5 (unless, of course, the research in question comes at the time within the scope of paragraph 5)"⁽⁴⁷⁾. Under such circumstances, the researching State or the international organization has an opportunity to modify the research project or to persuade the coastal State that the original research project will be carried out in accordance with the Convention.⁽⁴⁸⁾ If these endeavours fail, the coastal State is entitled to withhold consent.⁽⁴⁹⁾

(ii) Exclusively for Peaceful Purposes

The terms "peaceful purpose" or "peaceful uses" are both used in the LOS Convention. Only the Preamble of the LOS Convention and Article 301

---

⑮ LOS Convention Article 246 (3) provides that "[c]oastal States shall, in normal circumstances, grant their consent for marine scientific research projects by other States or competent international organizations in their exclusive economic zone or on their continental shelf to be carried out in accordance with this Convention exclusively for peaceful purposes and in order to increase scientific knowledge of the marine environment for the benefit of all mankind. To this end, coastal States shall establish rules and procedures ensuring that such consent will not be delayed or denied unreasonably".

⑯ See A. H. A. Soons, supra. note 116, at 165.

⑰ Ibid.

⑱ Ibid., at 166.

⑲ Ibid.

use the term "peaceful uses",[50] while Articles 88, 141, 143 (1), 147 (2) (d), 155 (2), 240 (a), and 246 (3) all use the term "peaceful purpose". However, there is not too much difference between these two terms.[51] Furthermore, there is no explicit definition of "peaceful purpose" in the Convention, whereas the meaning of "peaceful uses" seems to be defined by the LOS Convention Article 301, entitled "peaceful uses of the seas"[52]. It reads as follows:

> **In exercising their rights and performing their duties under this Convention, States Parties shall refrain from any threat or use of force against the territorial integrity or political independence of any State, or in any other manner inconsistent with the principles of international law embodied in the Charter of the United Nations.**

Article 88, relating to the high seas, provides that the high seas and the EEZ[53] are reserved for peaceful purpose. Articles 141, 143 (1), 147 (2) (d) and 155 (2) all concern the Area and provide that the Area shall be open to use exclusively for peaceful purposes. Article 240 (a) and 246 (3) deal with the regime of MSR and state that MSR shall be conducted exclusively for peaceful purposes.

---

[50] See Moritaka Hayashi, Military and Intelligence Gathering Activities in the EEZ: Definition of Key Terms, in: *Marine Policy*, Vol. 29, 2005, 123-137.

[51] Ibid., at 123. A further explanation could see Myron H. Nordquist (ed.), *United Nations Convention on the Law of the Sea 1982: A Commentary*, Vol. Ⅲ (Martinus Nijhoff Publishers, 1995), at 90. It is said that "Article 88 sets out the general principle that the high seas are to be used for peaceful purposes. In this respect, it echoes a theme first mentioned in the Preamble (see para. 88. 1 above). That principle is also set out in article 301, entitled 'peaceful uses of the seas,' [...]". It appears that the two terms are regarded as the same.

[52] See Boleslaw A. Boczek, Peaceful Purposes Provisions of the United Nations Convention on the Law of the Sea, in: *Ocean Development and International Law*, Vol. 20, 1989, 359-389, at 370.

[53] According to Article 58 (2), Article 88 also applies to the Exclusive Economic Zone.

Compared with other issues, "peaceful uses of the sea" did not gain too much attention at the third International Law of the Sea Conference.[354] A debate only took place at the fourth session of UNCLOS Ⅲ in the Plenary on the issue of "peaceful uses of ocean space: zones of peace and security".[355] Three trends regarding the interpretation emerged from the discussion on the peaceful uses of ocean space. At the 68th meeting,[356] the delegation of Iran summed up these opinions:

> Many States had taken the view that "peaceful purposes" meant the prohibition of all military activities, including activities by military personnel, on the sea-bed. Other States interpreted the principle as prohibiting all military activities for offensive purposes, but not, for instance, the use of military means of communication or the use of military personnel for scientific purposes. A third group of States maintained that the test of whether an activity was peaceful was whether it was consistent with the Charter of the United Nations and other obligations of international law.[357]

Subsequently, according to an anonymous proposal made in the eighth session in 1979, the "peaceful purposes concept was to be restricted to activities directed against the sovereignty, territorial integrity or political independence of other States",[358] despite the fact that the text was not included in LOS Convention Article 88 but was included as the new LOS Convention

---

[354] See Boleslaw A. Boczek, supra. note 352, at 368.

[355] See United Nations Conference on the Law of the Sea, Official Records, Vol. Ⅴ, 66th-68th Plenary Meeting, 54-71.

[356] Ibid., 63-68. Also see http://untreaty.un.org/cod/diplomaticconferences/lawofthesea-1982/docs/vol_V/a_conf-62_sr-68.pdf. (last visited on 15 April 2013).

[357] Ibid., at 65. Also see http://untreaty.un.org/cod/diplomaticconferences/lawofthesea-1982/docs/vol_V/a_conf-62_sr-68.pdf, at 65 (last visited on 15 April 2013).

[358] See Alexander Proelss, Peaceful Purpose, in: Wolfrum (ed.) *Max Plank Encyclopedia of Public International Law*, Vol. Ⅷ, para. 14, at 197.

Article 301.⁽⁵⁹⁾ It was further stipulated by the UN Secretary-General that "[m]ilitary activities which are consistent with the principle of international law embodied in the Charter of the United Nations, in particular with Article 2, paragraph 4 and Article 51, are not prohibited by the Convention on the Law of the Sea"⁽⁶⁰⁾. Bateman also notes that:

> In accordance with UNCLOS [LOS Convention] Article 58 and 88, the EEZ should be reserved for peaceful purposes. This of itself does not exclude military operations in the EEZ as the argument is well accepted that provided such operations are not aggressive, involving the threat or use of force against the coastal State, they are consistent with the UN Charter.⁽⁶¹⁾

With respect to the Area, the content of the peaceful purposes clause is even less clear than in the case of LOS Convention Article 88.⁽⁶²⁾ Although the first proposal drafted by Arvid Pardo in 1967 (so-called Malta Initiative) and the resolution submitted by fifteen States in 1970 aimed at a complete demilitarization of the international seabed area, from the beginning of the third session of UNCLOS Ⅲ in 1975, the language concerning the peaceful purposes clause corresponded to the one finally employed in LOS Convention Article 141.⁽⁶³⁾ As Proelss stated:

> It seems that with the conclusion of the Seabed Arms Control Treaty in 1971, the demand to widen the scope of application of Article 141 UN Convention on the Law of the Sea [LOS Convention] to military issues diminished in the framework of the Conference.⁽⁶⁴⁾

Therefore, the peaceful purposes clause was generally not to be

---

⁝⁹ See Alexander Proelss, Peaceful Purpose, in: Wolfrum (ed.) *Max Plank Encyclopedia of Public International Law*, Vol. Ⅷ, para. 14, at 197.

⁶⁰ Excerpted from Ibid., para. 15, at 197-198.

⁶¹ See Sam Bateman, supra. note 287, at 116.

⁶² See Alexander Proelss, supra. note 358, para. 18.

⁶³ Ibid.

⁶⁴ Ibid.

understood as requiring a complete demilitarization of the maritime zones concerned. As no internationally accepted definition exists, the clause is applied on a case-by-case basis.[365] In regards to MSR, the concept of research for exclusively peaceful purpose is not defined, but it cannot be "sweepingly" interpreted as "research for 'non-military' purpose".[366] Soons addresses this issue, proposing that "[i]t can therefore be concluded that marine scientific research may be conducted for military purposes, provided those purposes are not incompatible with the UN Charter"[367]. In other words, such research conducted for military purpose is generally subject to the same regime as that of ordinary MSR.[368]

(iii) Increase Scientific Knowledge of the Marine Environment for the Benefit of All Mankind

The task of MSR is to sufficiently understand the marine world in order to predict and explain the changes therein by means of observing, analysing and so on.[369] Scientific knowledge comes with scientific and technological developments, which can have profound effects on the world and human populations. Specifically, science and technology can be important triggers for legal developments.[370] From the doctrine of the cannon shot rule[371] to the appearance of the EEZ, it illustrates that technological capacity has become a major driver for the extension of the coastal State's jurisdiction.[372] Adoptions of new Conventions, due to the advent of new circumstances, are the outcomes of scientific and technological developments. Examples include the Convention for the Protection of Submarine Telegraph Cables, which was

---

[365] See Alexander Proelss, supra. note 358, para. 22.
[366] See Boleslaw A. Boczek, supra. note 352, at 376.
[367] See A. H. A. Soons, supra. note 116, at 135.
[368] Ibid.
[369] See Florian H. Th. Wegelein, supra. note 219, at 9.
[370] See Aldo Chircop, supra. note 27, at 579.
[371] It means the use of the range of coastal guns as a useful outer limit of the territorial sea.
[372] See Aldo Chircop, supra. note 27, at 578.

adopted in 1884 because of the laying of transoceanic submarine telegraph cables,⑬ as well as the Convention for the Unification of Certain Rules Respecting Collisions between Vessels which was adopted in 1910 because of the advent of the steam engine on board ships and metal hulls.⑭ The legal development process could be occasioned by even a rudimentary scientific understanding of the oceans, together with the awareness of its potential.⑮

Apart from what is analysed above, there is another angle to interpret "increase scientific knowledge of the marine environment for the benefit of all mankind", which could also lead back to the distinction between "fundamental" and "applied" scientific research. The bright spot of LOS Convention Article 246 is the requirement for coastal States to grant consent if the objective of the research project is to "increase scientific knowledge of the marine environment for the benefit of all mankind".⑯

P. K. Mukherjee points out that:

> [t]his appears to be an implicit recognition of the existence of some forms of oceanic research which are indeed fundamental in nature, and represents a latent revival of the distinction between pure and applied research which has been supposedly abandoned in the current text.⑰

On the other hand, Abdulaqwi A. Yusuf comments that:

> [i]f the criterion for ascertaining pure research is the language of [A]rticle 246 (3), that is, that the activity seeks to contribute to man's knowledge of the maritime environment for the benefit of all, then such an ascertainment inevitably becomes

---

⑬ See Aldo Chircop, supra. note 27, at 577.
⑭ Ibid.
⑮ Ibid., at 578.
⑯ See P. K. Mukherjee, supra. note 163, at 102.
⑰ Ibid.

79

highly subjective and imprecise.[178]

Some scholars have argued that the approaches used to differentiate between "fundamental" and "applied" research should be stopped. The real concern that focus should be put on is whether the research is beneficial for all mankind or not, and this should depend on beneficial effect. If there is an "effect", it should be immediately apparent or be at least foreseeable in the near future, in which case the research project should be promoted. Mukherjee also states that "[...] if the benefit of mankind as a whole is overwhelming, then regardless of the ultimate profit-oriented motive of the research [S]tate, if any, the freedom of research should not be hampered"[179].

(d) Discretion of the Coastal State to Withhold Consent

There are four specific circumstances under LOS Convention Article 246 (5) (a)—(d)[180] where a coastal State has the discretion to withhold consent for the conduct of MSR (where the research is conducted in the coastal State's EEZ or continental shelf). If the consent has been withheld, "the coastal State will have to indicate which subparagraph it considers

---

[178] See Abdulaqwi A. Yusuf, Toward a New Legal Framework for Marine Research: Coastal-State Consent and International Coordination, in: *Virginia Journal of International Law*, Vol. 19, 1979, 411-429, at 419.

[179] See P. K. Mukherjee, supra. note 163, at 113.

[180] LOS Convention Article 246 (5) provides that: "Coastal States may however in their discretion withhold their consent of the conduct of a marine scientific research project of another State or competent international organization in the exclusive economic zone or on the continental shelf of the coastal State if that project: (a) is of direct significance for the exploration and exploitation of natural resources, whether living or non-living; (b) involves drilling into the continental shelf, the use of explosives or the introduction of harmful substances into the marine environment; (c) involves the construction, operation or use of artificial islands, installations and structures referred to in Articles 60 and 80; (d) contains information communicated pursuant to Article 248 regarding the nature and objectives of the project which is inaccurate or if the researching State or competent international organization has outstanding obligations to the coastal State from a prior research project."

applicable. In case of a dispute, the burden of proof lies with the coastal State"[81].

(i) LOS Convention Article 246 (5): Subparagraph (a)

LOS Convention Article 246 (5) (a) provides that if the research project "is of direct significance for the exploration and exploitation of natural resources, whether living or non-living", the coastal State may on its discretion withhold consent. However, the formulation is not precise and would undoubtedly give rise to different interpretations.[82] There are two ambiguities that could arise in this clause. One ambiguity is the denotation of exploration and exploitation and the difference between them. The other ambiguity is the impact of MSR on the exploration and exploitation of natural resources. In this regard, the clause adopts a progressive method of expression: not only should the research be relevant to the exploration and exploitation of natural resources, but the relevance or significance must be direct as well.[83]

First, concerning the notion of "exploration and exploitation", Soons defined "exploration" as "data collection activities (scientific research) concerning nature resources, whether living or non-living, conducted specifically in view of the exploitation (i.e. economic utilization) of those natural resources"[84]. Clearly, Soons made the connection between "exploration" and "exploitation". The exploratory activities described here are different from the research, which aims at obtaining more scientific knowledge. "Exploration" is carried out with a clear view to subsequent exploitation, irrespective of whether or not the exploitation would happen.[85] It is expected that the research results could uncover the location of natural resources and, more importantly, would facilitate assessing and monitoring of

---

[81] See A. H. A. Soons, supra. note 116, at 170.
[82] Ibid.
[83] Ibid.
[84] Ibid., at 125.
[85] See Florian H. Th. Wegelein, supra. note 219, at 85.

the natural resources with respect to their status and availability for commercial exploitation.[86] In regards to the definition of "exploitation", it is characterized as "the act of taking advantage of something"[87]. It denotes that "exploitation" is clearly linked to the utilization of natural resources, not just their discovery.[88] "Exploration" is arguably the antecedent stage of "exploitation". However, the "exploration" does not necessarily lead to the "exploitation", whereas the "exploitation" requires the "exploration" as basis.

In respect to the impact of MSR on exploring and exploiting natural resources, Soons asserts: "[...] it can be said that no piece of scientific knowledge can be guaranteed 'pure', or free of application."[89] Scientific research in the oceans increases the general knowledge of resources, and even though it is indirect, it certainly has an effect on exploring or exploiting natural resources.[90] Soons suggests: "almost all the marine scientific research could be considered to have some significance for the exploration or exploitation of natural resources."[91] Obviously, LOS Convention Article 246 (5) (a) is not intended to cover all MSR.[92] Here, the point of reference is "resource exploitation oriented" research. Thus, the term "direct" is of great importance. What makes the term "direct" applicable is the research project's objective. Whenever the research project is clearly conducted with a view to exploring or exploiting natural resources, this denotes "direct significance". The discovery of a commercial-sized population of lobsters by the National Oceanic and Atmospheric Agency's scientists in the estuarine channels along the coast of Maine could illustrate a direct relationship

---

[86] See Florian H. Th. Wegelein, supra. note 219, at 84.

[87] See Bryan A. Garner (ed.), *Black's Law Dictionary*, 7th edition, (St. Paul, Minn., 1999), at 600.

[88] See Florian H. Th. Wegelein, supra. note 219, at 85.

[89] See A. H. A. Soons, supra. note 116, the Note to Introductory Remarks, note 31.

[90] See Herman T. Franssen, supra. note 96, at 158.

[91] See A. H. A. Soons, supra. note 116, at 171.

[92] Ibid.

between research and resource exploitation.[83] Similarly, Wegelein argued that the word "direct" qualifies the term "significance" for exploration and exploitation as to be limited "both in substance and in time".[84] On the one hand, "substance" would mean the relevance or intention of the research project for exploitation, substituting possible exploratory research.[85] "Time", on the other hand, means that the data, information or research results "must be sufficient to allow for exploitation in the foreseeable future with the technology available".[86]

However, it seems difficult to deny that scientific research has laid the foundations for commercial exploration and exploitation of ocean resources.[87] As previously mentioned, geologists intend to learn more about the earth's crust and seabed structure. But their surveys, samples, or research results can be used to indicate whether or not there is a potential for oil exploitation. In this case, resource exploration is not the primary concern; the research activities are prompted by the geologists' curiosity rather than a possible economic return.[88] This illustrates the indirect significance for exploration and exploitation. Thus, one must conclude that the coastal State can withhold its consent only under the circumstances that the research project, as analysed above, is of direct significance for exploring and exploiting the natural resources.

(ii) LOS Convention Article 246 (5): Subparagraph (b)

Two kinds of activities are referred to in subparagraph (b). One is the activity that involves drilling into the continental shelf, while the other is the activity that uses explosives or introduces harmful substances into the marine environment. Both activities are regulated not only by Part XIII but also by Part VI as well as Part XII of the LOS Convention, which shed

---

[83] See Herman T. Franssen, supra. note 96, at 158.
[84] See Florian H. Th. Wegelein, supra. note 219, at 87.
[85] Ibid.
[86] Ibid.
[87] See Herman T. Franssen, supra. note 96, at 159.
[88] Ibid.

additional light on how to interpret this paragraph.

With respect to the activity involving drilling into the continental shelf, based on the LOS Convention Part VI, Article 81, the coastal State has the exclusive right to authorize and regulate drilling on the continental shelf for all purposes. It should be noted that the word "exclusive" means any drilling into the continental shelf that is subject to the coastal State's complete control.[399] Considering that drilling into the continental shelf may be regarded as a method to conduct MSR, according to LOS Convention Articles 246 and 81, the coastal State not only enjoys the discretion to withhold consent, but exercises complete control over the activity also.

With respect to activities that use explosives or introduce harmful substances into the marine environment, the major concern is the protection of the marine environment. Explosives could cause marine pollution, which has serious deleterious effects on marine life, such as the destruction of the natural habitats of living organisms or the killing of certain species.[400] Introducing harmful substances into the marine environment can certainly be regarded as pollution,[401] which may affect marine and even human life.[402] Based on LOS Convention Article 192,[403] all States have the obligation to protect and preserve the marine environment. Meanwhile, according to LOS Convention Article 56 (1) (b) (iii), the coastal State enjoys jurisdiction regarding the protection and preservation of the marine environment in the

---

[399] See A. H. A. Soons, supra. note 116, at 173.

[400] Ibid., at 172.

[401] It is said in Article (1) (4) of the LOS Convention that "pollution of the marine environment" means the introduction by man, directly or indirectly, of substances or energy into the marine environment, including estuaries, which results or is likely to result in such deleterious effects as harm to living resources and marine life, hazards to human health, hindrance to marine activities, including fishing and other legitimate uses of the sea, impairment of quality for use of sea water and reduction of amenities.

[402] Ibid.

[403] LOS Convention Article 192 sets out the general obligation that: "States have the obligation to protect and preserve the marine environment."

EEZ. Thus, similar to the first category, under such circumstances the coastal State has, besides the discretion to withhold the consent, the right to take all necessary measures (consistent with the LOS Convention) to prevent, reduce, and control the pollution.[404]

(iii) LOS Convention Article 246 (5): Subparagraph (c)

According to subparagraph (c), the coastal State may withhold its consent to the conduct of MSR that involves the construction, operation, or use of artificial islands, installations, or structures, as referred to in LOS Convention Articles 60 and 80. As is well known, constructing artificial islands, installations, and structures on the seas for various purposes is a new trend to utilize the vast ocean space and resources.[405] According to LOS Convention Article 60 and Article 56 (1) (b) (i), the coastal State shall have the exclusive right to construct, authorize, and regulate the construction, operation, and use of artificial islands, installations, and structures in the EEZ for the purposes of MSR. Besides, it should be noted that from the wording of subparagraph (c), not only the construction of artificial islands, installations, and structures is subject to the exclusive jurisdiction of the coastal State, but the operation, and use of artificial islands, installations, and structures which already exist also fall within the ambit of this subparagraph. Namely, a "coastal State can withhold consent to a marine scientific research project even if that project merely involves the use of one or more already existing artificial islands, installations or structures"[406].

(iv) LOS Convention Article 246 (5): Subparagraph (d)

Two different cases are included in subparagraph (d). First, the coastal State can withhold consent if the information it received from the researching

---

[404] According to the LOS Convention Article 194 (1).
[405] See Nikos Papadakis, Artificial Islands, Installations and Structures in the Exclusive Economic Zone, in: *La Zona Economica Esclusiva* (Milano, 1983), 97-114, at 99.
[406] See A. H. A. Soons, supra. note 116, at 174.

State or the international organization is inaccurate, pursuant to LOS Convention Article 248, regarding the nature and objective of the research project. It may seem a little familiar, since LOS Convention Article 253 provides a similar reference. However, the difference lies in the fact that LOS Convention Article 246 (5) (d) only deals with the information "regarding the nature and objectives of the project", while Article 253 involves all the information pursuant to LOS Convention Article 248. In addition, the consequences are varied for Articles 246 (5) (d) and 253. Under Article 246 (5) (d), consent can be withheld at the very beginning, while according to Article 253, after the research project proceeds, suspension and cessation may be the consequences. Nonetheless, suspension is not the final result. It can be lifted by the coastal State once researching State or international organization complies with the conditions required under LOS Convention Articles 248 and 249 after being notificated of the suspension. This means that the researching State or international organization has the opportunity to rectify the situation within a reasonable time period. Still, the suspension can lead to a cessation if the rectification does not happen. In the light of LOS Convention Article 246 (5) (d), the information regarding the nature and objectives of the project is so crucial that it will probably lead to a withholding of consent to conduct the MSR project from the beginning. The issues of the research project's nature and objectives are also sensitive; they might also involve homeland security, resource exploration, exploitation, and some other issues. To avoid impairing to its rights and interests, the coastal State is entitled to withhold consent. The problem lies in determining whether the information is accurate. This depends on other information that the researching State or international organization provides, and to some extent on the coastal State's own scientific capabilities.[407] In this case, the coastal State's judgment may have subjective tendencies. Therefore, the criteria and guidelines provided in

---

[407] See A. H. A. Soons, supra. note 116, at 174.

LOS Convention Article 251[408] could play an important role.[409] With these criteria and guidelines, it would be easier for most coastal States to ascertain the research project's nature and objectives when they evaluate and verify the information provided to them pursuant to LOS Convention Article 248.[410] This is a better way to promote the accumulation of marine environment knowledge and in turn improve the world environment at large.

Second, the coastal State may also withhold consent if the researching State or international organization has outstanding obligations to the coastal State from a prior research project. The obligations referred to are included in LOS Convention Article 249.[411] It should be noted that the conditions listed in LOS Convention Article 249 "only apply to research projects which come within the scope of the duty of the coastal State to grant consent in

---

[408] LOS Convention Article 251 provides that: "States shall seek to promote through competent international organizations the establishment of general criteria and guidelines to assist States in ascertaining the nature and implications of marine scientific research."

[409] See A. H. A. Soons, supra. note 116, at 174.

[410] Ibid., at 196.

[411] See LOS Contention Article 249 provides the duty to comply with certain conditions. The conditions are: (a) to ensure the right of the coastal State, if it desires, to participate or be represented in the marine scientific research project, especially on board research vessels and other craft or scientific research installations, when practicable, without payment of any remuneration to the scientists of the coastal State and without obligation to contribute towards the costs of the project; (b) to provide the coastal State, at its request, with preliminary reports, as soon as practicable, and with the final results and conclusions after the completion of the research; (c) to undertake to provide access for the coastal State, at its request, to all data and samples derived from the marine scientific research project and likewise to furnish it with data which may be divided without detriment to their scientific value; (d) if requested, to provide the coastal State with an assessment of such data, samples and research results or provide assistance in their assessment or interpretation; (e) to ensure, subject to paragraph 2, that the research results are made international available through appropriate national or international channels, as soon as practicable; (f) to inform the coastal State immediately of any major change in the research programme; (g) unless otherwise agreed, to remove the scientific research installations or equipment once the research is completed.

normal circumstances according to paragraph 3 of Article 246"[12]. It means that in normal circumstances, the coastal State would grant its consent when the researching State and international organization comply with the conditions listed in LOS Convention Article 249. If some or all of the outstanding obligations are not complied with, the coastal State is entitled in its discretion to withhold the consent to a future application to conduct MSR in its EEZ or continental shelf. Soons pointed out that "[i]f the outstanding obligations are fulfilled after the coastal State has informed the researching State or international organization of its decision to withhold consent, a new application for consent will have to be made"[13].

(e) MSR on the Outer Continental Shelf

With the entry into force of the LOS Convention, the regime for the outer continental shelf was created.[14] Based on LOS Convention Article 76 (1), "a two-pronged definition of the continental shelf" is provided.[15] Regarding this "two-pronged" definition, it is expressed that "[o]ne [definition] encompasses the natural prolongation of the continental land mass to the outer edge of the continental margin and the second [definition] establishes a juridical shelf extending to a distance of 200 nautical miles from shore, irrespective of the existence of a naturally prolonged continental land mass"[16]. Whenever the outer edge of the continental margin extends beyond 200 nautical miles from the baselines that the breadth of the territorial sea is measured, the coastal State shall establish the outer limits of the margin by either one of the two methods accorded in LOS Convention Article 76 (4)

---

  [12] See A. H. A. Soons, supra. note 116, at 188.

  [13] Ibid., at 175-176.

  [14] See Ted L. McDorman, The Entry into Force of the 1982 LOS Convention and the Article 76 Outer Continental Shelf Regime, in: *The International Journal of Marine and Coastal Law*, Vol. 10, 1995, 165-187, at 185.

  [15] See Kilaparti Ramakrishna, Robert E. Bowen and Jack H. Archer, Outer Limits of Continental Shelf—A Legal Analysis of Chilean and Ecuadorian Island Claims and US response, in: *Marine Policy*, Vol. 11, 1987, 58-68, at 59.

  [16] Ibid.

(a) (i) and (ii). However, the outer limit lines created pursuant to LOS Convention Article 76 (4) are not extending beyond either 350 nautical miles from the baselines or 100 nautical miles from the 2500 meter isobaths. Apart from the technical criteria for establishing the continental shelf's outer limit, the outer continental shelf regime also includes the procedures that establish the outer limit (which would refer to the Commission on the Limits of the Continental Shelf set up under Annex B to the LOS Convention) and the revenue-sharing requirement for resources derived from the continental shelf beyond 200 nautical miles under LOS Convention Article 82. The purpose of the outer limit criteria is not only to determine the outer limit of the adjacent continental margin, but also to clearly demarcate the geographical domain of the International Seabed Authority.[417] Further, the revenue-sharing requirement can be seen as a necessary consequence of the recognition of national jurisdiction over the outer continental margin, which would attenuate the chance of the International Seabed Authority to have more control on the geographic area.[418]

Returning to LOS Convention Article 246 (6), it is stipulated that under certain circumstances, the coastal State may not exercise its discretion to withhold consent, even if the MSR undertaken on the continental shelf is of direct significance for the exploration and exploitation of natural resources, whether living or non-living. The prerequisites are:

(i) the **MSR project is conducted in accordance with provisions of LOS Convention Part XIII**;

(ii) **it is conducted beyond 200 nautical miles from the baselines from which the breadth of the territorial sea is measured**; and

(iii) **the site of the research project should be outside those specific areas which coastal State may at any time publicly**

---

[417] See Ted L. McDorman, supra. note 414, at 175.
[418] Ibid., at 174.

designate as areas in which exploitation or detailed exploratory operations focused on those areas are occurring or will occur within a reasonable period of time."[119]

However, this does not mean that the researching State or organization could be exempted from the consent regime. As a compromise, coastal States seek to extend their control over the outer continental shelf at the expense of impairing their discretion to withhold the consent. Still, MSR is subject to the consent regime. Seen from another angle, the freedom of MSR used to have a wider coverage. By virtue of LOS Convention Article 246 (6), in terms of those specific areas that the coastal State may at any time publicly designate as exploitation or detailed exploratory operations focused areas, the regime of the outer continental shelf and the provision in question "gives the coastal State the right to exclude certain area of the outer continental shelf from the freedom of scientific research under the condition that exploration and exploitation are carried out or will commence 'within a reasonable period of time'"[120]. However, this has led to interpretative ambiguity. The qualification of "detailed" is hard to determine since it depends on different circumstances, such as the available technology and the ability to make use of the technology.[121] Moreover, "exploratory operations" can include a number of operations, which would make the application of this provision very complicated.[122]

D. Conclusion

Marine science is the term applied to "the scientific investigation of the ocean, its biota and its physical boundaries with the solid earth and the at-

---

[119] See LOS Convention Article 246 (6).
[120] See Florian H. Th. Wegelein, supra. note 219, at 204.
[121] Ibid.
[122] Ibid.

mosphere"[123]. Politics surrounding the conduct of MSR are as intricate and complex as the methods, means and technologies employed in the conduct of its many facets. [124] The legal regime regulating the conduct of MSR activities has evolved throughout the centuries, from the seventh session of the ILC in 1955, to the adoption of LOS Convention in 1982.

Prior to the CCS 1958, there were no global instruments regulating the conduct of MSR. Until the adoption of CCS 1958, a consent regime regarding research activities on the continental shelf was established, despite of the fact that some of the wording problems had led the implementation of this provision subject to a certain degree of difficulties. In light of the LOS Convention, among the various legal regimes that it provides, the conduct of MSR has, for the first time in the history of the codification of the International Law of the Sea, acquired a prominent place within the new comprehensive regime for the oceans. [125] During the negotiating process of the LOS Convention, marine science and technology problems had been considered as an important and indispensable component of the overall "package" which is a genuine reflection of the ever-growing role of oceanic science and technology in all activities relating to the seas and their resources. [126] The LOS Convention has retained the basic principle of consent by coastal States for research on the continental shelf and extended it to the EEZ. [127] Furthermore, it has considerably expanded provisions governing

---

[123] Intergovernmental Oceanographic Commission (of UNESCO), Ocean Science for the Year 2000, 12th Session of the Assembly, Paris, Nov. 3-20 1982, in: *Ocean Yearbook*, Vol. 4, 1983, 176-259, at 181. Excerpted from Montserrat Gorina-Ysern, supra. note 192, Introduction, at xxii.

[124] See Montserrat Goriná-Ysern, ibid., at xxiv.

[125] See Alexander Yankov, A General Review of the New Convention on the Law of the Sea: Marine Science and Its Application, in: *Ocean Yearbook*, Vol. 4, 1983, 150-175, at 151.

[126] Ibid.

[127] See Office for Ocean Affairs and the Law of the Sea, The Law of the Sea: National Legislation, Regulations and Supplementary Documents on Marine Scientific Research in Areas under National Jurisdiction (United Nations, 1989), Introduction, at 1.

MSR by adding general principles as well as detailed implementation rules regarding its conduct.[128]

While this chapter has examined the entire legal regime regarding MSR provided in LOS Convention Part XIII, particular emphasis has been placed on Article 246, not only due to the fact that it simultaneously addresses the conduct of MSR in the EEZ and on the continental shelf, but also because of the complexity of the legal regime with respect to the EEZ on the one hand and the continental shelf beyond 200 miles from the baselines on the other. It is well known that the newly introduced concepts of the EEZ and the outer continental shelf had gone through fierce controversies during the negotiating process of LOS Convention. Moreover, regarding MSR, the most important issues, and those on which there was the greatest divergence of views, centred upon research in the EEZ as well.[129] Therefore, MSR conducted in the EEZ and on the continental shelf would require more detailed consideration as a main component of the MSR regime, also with regard to more recent technologies.

---

[128] See Office for Ocean Affairs and the Law of the Sea, The Law of the Sea: National Legislation, Regulations and Supplementary Documents on Marine Scientific Research in Areas under National Jurisdiction (United Nations, 1989), Introduction, at 1.

[129] See John R. Stevenson, Bernard H. Oxman, The Third United Nations Conference on the Law of the Sea: The 1974 Caracas Session, in: *American Journal of International Law*, Vol. 69, 1975, 1-30, at 28.

III. Implementing the MSR Regime in the Marine High-Tech Era

Over the last 30 years, marine science has changed remarkably in response to a better understanding of problems such as climate change, sea level rise, and dwindling fisheries. From a spatial perspective, the modern study of the marine environment is significantly wider in scope since it includes the study of the air and atmosphere above the ocean, and terrestrial and hydrospace, more generally, as a whole.[430] The exclusion of a MSR definition enabled the negotiators to reach a consensus on the draft of LOS Convention Part XIII. While the absence of a definition poses difficulties for a full identification of those MSR activities that fall within or outside the MSR regime, it enables flexibility for adapting the spirit of Part XIII to today's new research frontiers and opportunities due to the undefined boundary for MSR as a concept.[431] From a present-day point of view, many rights and duties in the LOS Convention were negotiated on the basis of outdated marine scientific knowledge and technology. It is stated that the LOS Convention has consequently become inadequate in responding to changing demands of the international society, leaving further legal development to state practice and evolving customary international law.[432] Whether this is correct or not will be put to a test in the following by addressing three case studies that have newly emerged in the era of marine high-tech.

---

[430] See Aldo Chircop, supra. note 27, at 599.
[431] Ibid., at 600.
[432] Ibid., at 580.

A. Ocean Upwelling

1. Introduction

There is growing interest in using flap-valve operated ocean pipes to upwell nutrient-rich deeper waters in order to fertilize the surface ocean.⑬ Pumping up cold deep seawater into less fertile waters at the surface (ocean upwelling) may result in a significant enhancement of biological production, sequestration of atmospheric CO_2 and lowering the sea surface temperature.⑭ It has been suggested that "artificial upwelling on a very large scale in areas of the ocean where typhoons form could cool the surface just enough to prevent them, or at least reduce their severity"⑮. In this respect, ocean upwelling is one component of a portfolio of controversial technologies summarized under the heading climate engineering.⑯ These technologies aim at contributing to the fulfillment of the objective of the United Nations Framework Convention on Climate Change (UNFCCC) to "achieve [...] stabilization of greenhouse gas concentrations in the atmosphere at a level that would prevent dangerous anthropogenic interference with the climate system"⑰ by manipulating the global climate

---

⑬ See Andreas Oschlies, et al., Climate Engineering by Artificial Ocean Upwelling: Channelling the Sorcerer's Apprentice, in: *Geophysical Research Letters*, Vol. 37, 2010, 1-5, at 1.

⑭ Ibid., at 3.

⑮ See Brian Kirke, Enhancing Fish Stocks with Wave-Powered Artificial Upwelling, in: *Ocean and Coastal Management*, Vol. 46, 2003, 901-905, at 903.

⑯ The analysis regarding Ocean Upwelling is based on the article: Alexander Proelss, Hong Chang, Ocean Upwelling and International Law, in: *Ocean Development and International Law*, Vol. 43, Issue 4, 2012, at 371-385.

⑰ United Nations Framework Convention on Climate Change of 9 May 1992, Article 2, 1771 U. N. T. S. 107.

III. Implementing the MSR Regime in the Marine High-Tech Era

system either through interventions in the global carbon cycle (carbon dioxide removal-CDR) or by shielding solar radiation (solar radiation management-SRM).[438]

The conceptual idea is that the enhancement of nutrients will, if and to the extent to which the appropriate ocean environment is available,[439] generate phytoplankton blooms, which will lead to an enhanced absorption of dissolved CO_2 and to the generation of oxygen through the process of photosynthesis. When the phytoplankton is consumed by higher trophic levels, or when the phytoplankton dies, some of the carbon-containing organic matter will sink into the ocean interior before being remineralized back to CO_2. Depending on the depth and region of remineralization, this CO_2 will be out of contact with the atmosphere from decades to many centuries.[440]

The potential of ocean upwelling is not assessed exclusively for the purpose of CDR. In Japan, it is intended that an ocean farm called "Laputa Project" will be established in the near future.[441] "Laputa" was first described by Jonathan Swift as a floating island in *Gulliver's Travels*. This project was initiated in order to enhance ocean fish production by cultivating phytoplankton.[442] If upwelling pipes can be deployed on a large scale, it is anticipated that "an ocean farm and consequently a green floating island can

---

[438] See the definition attached in a note to Decision X/33, "Biological Diversity and Climate Change", of the Conference of the Parties to the Convention on Biological Diversity, available at: www.cbd.int/decision/cop/? id=12299, (last visited on 15 April 2013).

[439] See Andreas Oschlies, supra. note 433, at 1.

[440] See Ken Caldeira, et al., Depth, Radiocarbon, and the Effectiveness of Direct CO_2 Injection as an Ocean Carbon Sequestration Strategy, in: *Geophysical Research Letters*, Vol. 29, 2002, 1-4.

[441] See Shigenao Maruyama, Artificial Upwelling of Deep Seawater Using the Perpetual Salt Fountain for Cultivation of Ocean Desert, in: *Journal of Oceanography*, Vol. 60, 2004, 563-568, at 563.

[442] See Brian Kirke, supra. note 435, at 902.

be realized in an ocean desert"[43]. The Japanese Fisheries Agency and the Marino-Forum 21[44] have since 2000 funded a long-term project dedicated to increasing primary production in order to make new fishing grounds available by upwelling deep ocean water. [45]

There are two approaches to introduce ocean pipes into the marine environment: the pipes may either be used in a free-floating manner or they may be moored to the seabed. [46] Given that mooring pipes to the seabed would require, depending on the individual ocean depths, up to several kilometers of cable as well as a vessel equipped with a high performing crane, the costs of such pipes will be significantly higher than those of free-floating pipes. Given also that ocean upwelling and its effects are surrounded by a considerable degree of scientific uncertainty, with research and testing still underway, the deployment of pipes has not yet reached any large scale or commercial level.

The purpose of the following analysis is to discuss the international legal aspects involved in the deployment and use of artificial ocean upwelling pipes and to offer some suggestions with respect to their future regulation. The central questions to be addressed are whether legal rules exist that govern the deployment of ocean pipes, and which States are entitled to exercise jurisdiction over these objects. [47] As the employment of ocean pipes

---

[43]   See Shigenao Maruyama, supra. note 441, at 564.

[44]   The Marino-Forum 21 is an industry and government organization forum which acts as a subsidiary of the Japanese Agency of Fisheries. Its mission is to promote development of fisheries within Japan's exclusive economic zone (EEZ). See Nai-Kuang Liang and Hai-Kuen Peng, A Study of Air-Lift Artificial Upwelling, in: *Ocean Engineering*, Vol. 32, 2005, 731-745.

[45]   See Kazuyuki Ouchi, Outline of the Ocean Nutrient Enhancer "TAKUMI", in: *Journal of Ocean Science and Technology*, Vol. 2, 2005, 9-15, at 9.

[46]   James E. Lovelock and Chris G. Rapley, "Ocean pipes could help the Earth to cure itself", available at: www.nature.com/nature/journal/v449/n7161/full/449403a.html (last visited on 15 April 2013).

[47]   The legal relevance of potential unintended ecological responses connected to the deployment of ocean pipes will not be dealt with in this article.

constitutes an ocean use which affects both maritime areas within and beyond the limits of national jurisdiction, the answers to these questions need to be addressed on the basis of the LOS Convention.

2. Legal Analysis

The LOS Convention was concluded, according to its Preamble, "to settle, in a spirit of mutual understanding and cooperation, all issues relating to the law of the sea". LOS Convention Part XIII, section 4 contains five articles (Articles 258 to 262), which deal with scientific research installations or equipment in the marine environment and are, thus, potentially applicable to the issue here. However, it is only necessary to ascertain whether ocean pipes ought to be considered as installations or equipment in terms of these provisions if their deployment would constitute MSR.

(a) Legal Classification of Ocean Upwelling

Although the LOS Convention contains a separate part dealing with MSR (Part XIII), it does not, as stated above,[448] offer any definition of the concept. According to a definition suggested in the legal literature, "research comprises of creative work undertaken on a systematic basis in order to increase the stock of knowledge, including knowledge of man, culture and society".[449] With respect to the particular case of MSR, this would require investigating processes within the marine environment, which is generally understood "to cover the seabed and subsoil thereof, the water column, and the atmosphere immediately above the water".[450] In the absence of an authoritative legal definition, MSR can, therefore, be understood as any study and experimental work designed to increase human knowledge of

---

[448] See supra. Chapter I, A.
[449] See Florian H. Th. Wegelein, supra. note 219, at 11.
[450] See A. H. A. Soons, supra. note 116, at 124. See also Katharina Bork., The Legal Regulation of Floats and Gliders—In Quest of a New Regime?, in: *Ocean Development and International Law*, Vol. 39, 2008, 298-328, at 303.

the seabed or the subsoil, the water column, or the atmosphere directly above the water.[61] Furthermore, the idea of proper scientific attributes, being prerequisite for the legitimate scientific research that was brought forward for accessing the ocean fertilization activities, as referred to before, would benefit for enucleating the definition of MSR. The major criteria of proper scientific attributes have been identified above[62] and can be recapitulated as follows: (i) designed to answer questions that will add to the body of scientific knowledge; (ii) the design, conduct and/or outcomes of the proposed MSR activity should not be influenced by the economic interests; (iii) should be subject to scientific peer review at appropriate stages in the assessment process; (iv) should make a commitment to publish the results in peer reviewed scientific publications and include a plan in the proposal to make the data and outcomes publicly available in a specified time-frame.

As mentioned above, upwelling pipes are envisaged to enhance biological production, sequester atmospheric CO_2 and lower the sea surface temperature. Given these purposes, it is difficult to regard ocean upwelling as MSR, since the primary goal of CDR, as well as the development of fisheries, is not to increase human knowledge of the marine environment. Having said that, obtaining knowledge of the marine environment constitutes the initial stage of both kinds of activities (ocean upwelling and MSR) as seawater temperature, density, ingredient of nutrients and water currents are investigated at the proposed sites.[63] Thus, it is submitted that while the deployment of upwelling pipes for CDR or enhancement of fishery production on a large and/or commercial scale can not be regarded as MSR,

---

[61] See A. H. A. Soons, supra. note 116, at 124.
[62] See supra. Chapter I, A.
[63] See Seiko Ogiwara, etc., Conceptual Design of a Deep Ocean Water Upwelling Structure for Development of Fisheries, in: Proceedings of the Fourth (2001) Ocean Mining Symposium, (International Society of Offshore and Polar Engineers, 2001), at 150-157.

assessing the preconditions for ocean upwelling as well as testing artificial pipes does arguably meet the requirements of MSR.[464]

Different to applied MSR, basic MSR aims at acquiring new knowledge of the underlying foundation of phenomena and observable facts, without any particular application or use in view.[465] Its particular features are "openness, dissemination of data, free exchange of privately owned samples among researchers (in some cases), as well as publication and dissemination of research results"[466]. Consequently, basic MSR can be regarded as open research, which is intended for the benefit of all humankind, and that is characterized by prompt availability and full publication of the results.[467] Applied MSR, on the other hand, inter alia, is predominantly associated with the exploitation of resources[468] or aimed primarily at uncovering commercially useful information.[469] It is undertaken for specific practical purposes.[460] Due to its commercial relevance, there may be restrictions on the publication and the dissemination of the research results,[461] as proprietary issues may be involved.[462] The difference between basic and most of applied

---

[464] See generally Donald R. Rothwell, Tim Stephens, supra. note 37, at 321.

[465] Organization of Economic Cooperation and Development (OECD), *The Measurement of Scientific and Technological Activities: Proposed Standard Practice for Surveys of Research and Experimental Development* (OECD Publications Service, 1994), at 13.

[466] See Lyle Glowka, The Deepest of Ironies: Genetic Resources, Marine Scientific Research, and the Area, in: *Ocean Yearbook*, Vol. 12, 1996, at 154-178. See also Herman T. Franssen, supra. note 96, at 158-168.

[467] See John A. Knauss, supra. note 93, at 109. See also Herman T. Franssen, supra. note 96.

[468] See Florian H. Th. Wegelein, supra. note 219, at 70.

[469] See Lyle Glowka, supra note 456, at 172.

[460] See A. H. A. Soons, supra. note 116, at 6.

[461] See Florian H. Th. Wegelein, supra. note 219, at 70. See also W. Burger, Treaty Provisions Concerning Marine Science Research, in: *Ocean Development and International Law Journal*, Vol. 1, 1973, 159-184, at 170-171.

[462] See Lyle Glowka, supra note 456, at 173.

MSR has been summarized as follows:

> Intentions of an institution or of individuals claiming to conduct marine scientific research can be ascertained by examining whether the open publication of the results of the project is intended or not [...]. Neither exploration or exploitation activities nor resource-related or military research will meet the condition of open publication, for the results of such activities or research will necessarily remain secret; there is no reason, on the other hand, to refuse to publish the results of fundamental research.[63]

It should be noted that, in principle, the notion of MSR contained in the LOS Convention covers both kinds of scientific research.[64] As regards ocean upwelling, it is submitted that the deployment and testing during the "MSR phase" ought to be understood as applied MSR, since these activities are intended for specific practical purposes, either for climate engineering or natural resources exploitation. The jurisdictional implications of this conclusion will be dealt with below.

(b) Ocean Pipes as Scientific Research Installations or Equipment

Having ascertained that the use of ocean pipes in the testing phase constitutes MSR, it is mandatory to clarify the meaning of the terms "installations" and "equipment" as used in LOS Convention Articles 258-262, neither of which is, however, defined in the Convention. According to an ordinary understanding of the text,[65] "[i]nstallations are the larger types of

---

[63] See Lucius Caflisch, Jacques Piccard, The Legal Regime of Marine Scientific Research and the Third United Nations Conference on the Law of the Sea, in: *Zeitschrift für ausländisches öffentliches Recht und Völkerrecht*, Vol. 38, 1978, 848-901, at 850-851.

[64] See LOS Convention Articles 246 (3) and (5). See also A. H. A. Soons, supra. note 116, at 6.

[65] According to Article 31 (1) of the 1969 Vienna Convention on the Law of Treaties, 1155 U. N. T. S. 331, "[a] treaty shall be interpreted in good faith in accordance with the ordinary meaning to be given to the terms of the treaty in their context and in the light of its object and purpose".

the devices (not being ships) used for conducting scientific research; they include in any case all devices which are capable of being manned"[466]. Contrary to the term "installations", "[t]he term 'equipment' would then cover all other (smaller) research devices which are used separately from ships, for example instrument packages"[467]. Accordingly, one of the central differences between "installations" and "equipment" is the size of the objects concerned. This reading is supported by LOS Convention Article 260 according to which safety zones may be established around scientific research installations. By not making any reference to scientific research equipment, this provision clarifies that "installations" pose a greater danger to international navigation or other uses of the sea, which necessarily implies a certain size of the objects.[468] A further point of reference is the duration of deployment. It has been elaborated "while installations are intended to remain in place for an extended period of time or even permanent, equipment has the connotation of being quickly deployed and removed in the course of a single experiment".[469] Finally, in addition to the size and duration of the deployment, the intended functions are also to be regarded as one of the relevant factors. While the word "equipment" usually denotes objects used for a specific purpose or activity,[470] installations "[m]ay consist of a number of various parts with a variety of functions".[471]

With regard to ocean pipes to be used for artificial upwelling, size, time and purpose of the deployment all meet the criteria of equipment. Ocean

---

[466] See A. H. A. Soons, supra. note 116, at 235.
[467] Ibid.
[468] Ibid. See also Katharina Bork, *Der Rechtsstatus von Unbemannten Ozeanographischen Messplattformen im Internationalen Seerecht* (Baden-Baden, 2011), at 66-67.
[469] See Florian H. Th. Wegelein, supra. note 219, at 138.
[470] See Bryan A. Garner, supra. note 387, at 558.
[471] See Florian H. Th. Wegelein, supra. note 219, at 138.

pipes are comparatively small objects, with an average diameter of one meter. [412] Their length depends on where they are deployed and on the depth where cold, nutrient-rich water is available. Most of the pipes are made of plastic; their life span is likely to expire within weeks after deployment. [413] Their single function is to pump cold, nutrient-rich water to the ocean surface. It can be concluded that ocean upwelling pipes, as long as their use has not entered the commercial deployment phase, are to be considered as MSR equipment.

LOS Convention Articles 260-262 specifically address the relationship between marine scientific equipment and installations on the one hand and other legitimate uses of the oceans on the other. While the right to establish safety zones only exists in respect of installations, Article 261 emphasizes the requirement of non-interference with shipping routes in general terms. Article 261 states that: "[t]he deployment and use of any type of scientific research installations or equipment shall not constitute an obstacle to established international shipping routes." This provision clearly corresponds with LOS Contention Article 240 (c) according to which MSR is not to unjustifiably interfere with other legitimate uses of the sea compatible with the Convention.

Furthermore, LOS Convention Article 262 provides that installations or equipment are to bear identification markings indicating the State of registry or international organization to which they belong, and warning signals. This suggests that scientific research installations and equipment, similar to vessels, have to be registered with a State or an international organization. [414] With regard to ships, the LOS Convention provides that "[e]very State shall

---

  [412] See Andreas Oschlies, supra. note 433. Also see Shigenao Maruyama, supra. note 441.

  [413] Estimation based on experiences made in the context of the Ocean Productivity Perturbation Experiment (OPPEX). See Angelicque White, et al., An Open Ocean Trial of Controlled Upwelling Using Wave Pump Technology, in: *Journal of Atmospheric and Oceanic Technology*, Vol. 27, 2010, 385-396, at 390.

  [414] See Florian H. Th. Wegelein, supra. note 219, at 148.

III. Implementing the MSR Regime in the Marine High-Tech Era

fix the conditions for the grant of its nationality to ships, for the registration of ships in its territory, and for the right to fly its flag"[415]. Although the same provision also states that ships have the nationality of the State whose flag they are entitled to fly, and that there must exist a genuine link between the State and the ship, the International Tribunal on the Law of the Sea (ITLOS) clarified in the M/V Grand Prince Case that if the domestic law of a State provides that the right of a ship to fly its flag is directly connected to the act of registration of the ship in that State, registration is the decisive factor with regard to the nationality of the ship.[416] The nationality of a ship not only determines the rights and obligations to which it is subjected, but also reflects which State is responsible for an act or omission of the vessel, and which State is entitled to exercise diplomatic protection on its behalf.[417] In light of the above, identification markings could be regarded as a symbol of registration in the case of scientific research installations and equipment, which would indicate jurisdiction over the pipes. As one writer has stated:

> **The relevance of Article 262 lies in its context: installations [and equipment] used in marine scientific research operations must be attributable to a certain research project or State in order to determine whether or not they have been lawfully emplaced and do not constitute debris; also markings are necessary for the return to the rightful owner in case of loss; and finally, liability may be imposed on the basis of Article 262 markings.**[418]

Warning signals, according to LOS Convention Article 262, contribute to the safety at sea and of air navigation. The inclusion of the reference to

---

[415] LOS Convention Article 91 (1).

[416] See The M/V "Grand Prince" (Belize v. France), ITLOS Reports 2001, 17, paras. 83 et seq.

[417] See R. R. Churchill and A. V. Lowe, supra. note 91, at 209. See also Article 94, LOS Convention.

[418] See Florian H. Th. Wegelein, supra. note 219, at 150.

the safety of air navigation presumably references the fact that "certain installations could in certain locations present a danger to aviation due to their height"[79]. Since ocean upwelling pipes cannot realistically cause harm to air navigation, they cannot be considered as installations. Notwithstanding this, ocean upwelling pipes should, for the sake of safe navigation at sea, bear warning signals.

(c) Jurisdiction over Ocean Pipes and Deployment Requirements

The question, which State is entitled to exercise jurisdiction over ocean upwelling pipes, depends on where these are located and whether they are used for MSR, for climate engineering, or for the exploration and exploitation of the natural resources.

(i) Territorial Sea

According to LOS Convention Article 2, the coastal State enjoys sovereignty over its adjacent territorial sea, which is, however, subject to the regime of innocent passage laid down in Article 18. The innocent passage regime is restricted to foreign ships and, therefore, is not applicable vis-à-vis ocean upwelling pipes. Consequently, with regard to the installations or equipment relating to upwelling pipes, if these are located within a coastal State's territorial sea, the coastal State is entitled to exercise jurisdiction over them,[80] regardless of for what they are actually used. Similar to the situation in the internal waters, the jurisdiction of the coastal State enjoys priority over the jurisdiction of the registry State of the installations or equipment.[81]

In light of its sovereignty over the territorial sea, the coastal State is generally free to deploy ocean pipes. Having said that, the freedom of deployment is limited insofar as the coastal State is, if the pipes are used for MSR, obliged to follow the rules and principles contained in the general requirements concerning MSR (LOS Convention Articles 238 to 241) as well

---

[79] See A. H. A. Soons, supra. note 116, at 238.
[80] Ibid., at 234.
[81] Ibid.

as those codified in Articles 260 to 262. Additionally, given that the coastal State "shall not hamper the innocent passage of foreign ships through the territorial sea [...],"[82] the deployment of ocean pipes, irrespective of their purpose, is not to interfere with the right of innocent passage. In this respect, the coastal State is, arguably, entitled to request foreign ships to avoid certain areas of its territorial sea where ocean pipes have been deployed, provided that this measure does not make innocent passage impossible, or hamper innocent passage in an unjustifiable manner respectively. This is based on LOS Convention Article 21 (1) according to which: "[t]he coastal State may adopt laws and regulations, in conformity with the provisions of this Convention and other rules of international law, relating to innocent passage through the territorial sea, in respect of [...] (b) the protection of navigational aids and facilities and other facilities or installations; [...] (g) marine scientific research and hydrographic surveys [...]."[83]

(ii) EEZ

Pursuant to LOS Convention Article 56 (1) (a), the coastal State enjoys sovereign rights in its exclusive economic zone (EEZ) for the purpose of "exploring and exploiting, conserving and managing the natural resources" and for the "economic exploitation and exploration of the zone, such as the production of energy from the water, currents and winds". Prima facie, the fact that testing and deployment of ocean upwelling pipes involves some kind of exploration of ocean energy and living organisms favors the activity falling within the ambit of the sovereign rights of the

---

[82] See LOS Convention Article 24 (1).

[83] Admittedly, the scope of the coastal State's competence is not completely beyond doubt, as Article 22 (1), LOS Convention, supra note 15, authorizes the coastal State to "require foreign ships exercising the right of innocent passage through its territorial sea to use such sea lanes and traffic separation schemes as it may designate or prescribe for the regulation of the passage of ships" only "where necessary having regard to the safety of navigation", but not where necessary for the protection of facilities or installations.

coastal State. On closer inspection, however, the wording of LOS Convention Article 56 (1) (a), "for the purpose of" indicates that the purpose of the deployment, rather than the means, is to be regarded as the decisive factor for assessing whether or not the sovereign rights of the coastal State are affected or engaged. As demonstrated above, the deployment of ocean upwelling pipes may, depending on the circumstances, serve three different purposes.

Where the pipes are used for MSR, the coastal State may, according to LOS Convention Article 56 (1) (b) (i), lawfully exercise jurisdiction with regard to the establishment and use of artificial islands, installations and structures. This jurisdiction is further substantiated by LOS Convention Article 60 (1) according to which the coastal State has the exclusive right to construct and to authorize and regulate the construction, operation and use of artificial islands, installations and structures. It has been concluded that "exclusive" in terms of the provision means that "the emplacing State may not regulate even if the coastal State has not taken any regulatory measures"[44].

However, these provisions may only be relied upon as a basis for coastal State jurisdiction if ocean pipes qualify as installations or structures. As demonstrated above, it is doubtful that the question can be answered in the affirmative. It is submitted that the criteria identified above governing the differentiation between installations and equipment in the context of LOS Convention Article 258 et seq. (size, duration of deployment, function) are applicable also to LOS Convention Articles 56 and 60,[45] given that both LOS Convention Articles 60 (4) and 260 entitle the coastal State to establish safety zones around these objects in order to ensure the safety both of navigation and of the objects. Further evidence for this interpretation can be

---

[44] See Florian H. Th. Wegelein, supra. note 219, at 150.

[45] See Katharina Bork, supra. note 468, at 66 et seq. Also see Elmar Rauch, Military Uses of the Ocean, in: *German Yearbook of International Law*, Vol. 28, 1985, 229-258, at 255 et seq.

drawn from the trauvaix préparatores of the Convention. On the eve of UNCLOS Ⅲ, it was argued by the United States in the Committee on the Peaceful Uses of the Sea-Bed and the Ocean Floor Beyond the Limits of National Jurisdiction that "[f]or the purpose of this chapter, the term 'installation' refers to all off-shore facilities, installations, or devices other than those which are mobile in their normal mode of operation at sea"[86].

If this line of argument is followed, ocean pipes used for MSR do not qualify as installations or structures and the coastal State is not entitled to exercise jurisdiction over these objects on the basis of LOS Convention Article 56 (1) (b) (i) in conjunction with Article 60. In this case, the jurisdiction of the coastal State would stem from LOS Convention Article 56 (i) (b) (ii) in conjunction with LOS Convention Article 246. Consequently, the coastal State has the right to regulate, authorize and conduct the deployment of ocean pipes used for MSR in its EEZ or on its continental shelf.[87] It should be noted, however, that the scope of the discretion of the coastal State that exists in respect of whether or not it grants its consent to third State activities is limited by LOS Convention Article 246 (3), where it provides that:

> [c]oastal States shall, in normal circumstances, grant their consent for marine scientific research projects by other States or competent international organizations in their exclusive economic zone or on their continental shelf to be carried out in accordance with this Convention exclusively for peaceful purposes and in order to increase scientific knowledge of the marine environment for the benefit of all mankind. [...]

This limitation applies if and to the extent to which the deployment of ocean pipes is not of direct significance to the exploration and exploitation of

---

[86] UN Doc. A/AC. 138/SC. II/L. 35, 16 July 1973, reprinted in Reports of the Committee on the Peaceful Uses of the Sea-Bed and the Ocean Floor Beyond the Limits of National Jurisdiction, Vol. Ⅲ (1973), at 75.

[87] See LOS Convention Article 246.

natural resources, or touches upon any other of the motives mentioned in LOS Convention Article 246 (5). In contrast, the State which deploys the pipes is entitled to regulate their emplacement and operation and it is the addressee of the rules governing State responsibility and liability in case damage occurs.

Where the ocean pipes are used for the purpose of enhancing the fish productivity, LOS Convention Article 56 (1) (a) indicates that only the coastal State is entitled to decide upon and regulate their deployment.

Finally, where the deployment of ocean upwelling pipes is for climate engineering purposes, arguably, this is not covered by the coastal State's sovereign rights and jurisdiction in terms of LOS Convention Article 56 (1), nor by the freedoms of other States referenced in LOS Convention Article 58 (1). As mentioned above, the primary purpose of the use of ocean pipes for climate engineering can neither be seen as increasing human knowledge of the marine environment [LOS Convention Article 56 (1) (b) (ii)], nor as economically exploiting and exploring the EEZ and its resources [LOS Convention Article 56 (1) (a)] or the laying of pipelines and submarine cables [Article 58 (1)], but rather as removing carbon dioxide from the atmosphere in order to contribute to combatting global warming.[48] In this case, LOS Convention Article 59 is applicable according to which any conflicts that may arise between the interests of the coastal State and any other State or States in regard to the EEZ should be resolved "on the basis of equity and in the light of all the relevant circumstances, taking into account the respective importance of the interests involved to the parties as well as to the international community as a whole". This provision covers economic uses not included in LOS Convention Articles 56 (1) and 58 (1) as well as other non-economic uses of the EEZ, such as, e. g., the operation of ocean

---

[48] Even if the use of ocean upwelling pipes for climate engineering purposes should be conducted on a commercial basis, it is submitted that this would still not constitute an activity that could be compared with energy production from the water, currents and winds due to the missing functional link with the resources of the EEZ.

data acquisition systems.[489] Given that LOS Convention Article 59 does not assign sovereign rights or jurisdiction but rather is a conflict rule, the provision implicitly contains the message that activities covered by its terms are, in absence of a user conflict, to be considered as lawful. It follows that the deployment of ocean pipes for the purpose of climate engineering in a State's EEZ is not subject to the jurisdiction of the coastal State but ought to be done in a way that takes due regard to the interests of other States. This is supported by the view of LOS Convention Article 59 by one noted authority that:

> **Given the functional nature of the EEZ, where economic interests are the principal concern this formula would normally favor the coastal State. Where conflicts arise on issues not involving the exploration for and exploitation of resources, the formula would tend to favor the interests of other States or of the international community as a whole.**[490]

If one follows this line of argument that reflects the shift of emphasis in favor of the coastal State on which the regime of the EEZ is, arguably, based,[491] conducting climate engineering projects in the EEZ is essentially to be dealt with in a similar manner as on the high seas, where LOS Convention Article 87 (2) requires all States to exercise their freedoms with due regard for the interests of other States in their exercise of the freedom of the high seas. The potential consequences deriving from this conclusion will be discussed in the following section.

(ⅲ) High Seas

While LOS Convention Article 87 (1) does not expressly refer to ocean pipes, the provision clarifies that freedom of scientific research is part of the

---

[489] See R. R. Churchill and A. V. Lowe, supra. note 91, at 414. See also Katharina Bork, supra. note 468, at 105-106.

[490] See Myron H. Nordquist (ed.), *United Nations Convention on the Law of the Sea* 1982: *A Commentary*, Vol. Ⅱ (Martinus Nijhoff Publishers, 1993), at 569.

[491] See Alexander Proelss, supra. note 84, at 83 et seq.

freedom of the high seas. As regards the other two potential purposes of these objects identified above, the words "inter alia" in paragraph 1 indicate that "the list of specific freedoms in paragraphs 1 (a) to (f) is not exhaustive, and that the freedom of the high seas may entail more than the enumerated activities"[62]. Given that LOS Convention Article 87 affirms the rule that the high seas are open to all States, one must conclude that in principle all States have the right to deploy equipment such as ocean upwelling pipes in the high seas, as long as the deployment is exercised under the conditions laid down in the LOS Convention and by other rules of international law.

As regards the MSR purpose of ocean upwelling pipes, notwithstanding the fact that LOS Convention Article 87 (1) (f) stipulates that all States are entitled to rely on the freedom of scientific research on the high seas, this right has to be exercised in conformity with LOS Convention Parts VI and XIII. MSR on the outer continental shelf of a State, i.e., on a continental shelf that extends beyond 200 nautical miles is subject to the coastal State's jurisdiction insofar as, according to Article 246 (6), the coastal State has the right to exercise its discretion to withhold consent to the conduct of MSR by another State concerning areas which it has designated for exploration and exploitation of natural resources. The rights of the coastal State over the continental shelf beyond 200 nautical miles do not affect the legal status of the superjacent waters,[63] which belongs to the high seas. LOS Convention Article 257 clearly states that there is a right to conduct MSR "in the water column beyond the limits of the exclusive economic zone". The freedom to conduct MSR is thus inapplicable only if and to the extent to which the regime of the high seas (which, as a matter of principle, also covers the

---

  [62] See Myron H. Nordquist (ed.), *United Nations Convention on the Law of the Sea 1982: A Commentary*, Vol. III, supra. note 351, at 84.
  [63] See LOS Convention Article 78.

deep seabed)[94] is superseded by special rules to the contrary such as those contained in LOS Convention Part XI on the regime of the Area[95] or in Article 76 in conjunction with Article 246 (6) respectively.

The above applies only to the use of ocean upwelling pipes for MSR on the high seas. The legal situation, however, is not that different if the pipes are used for the exploitation of natural resources or climate engineering. As regards the natural resources aspect, LOS Convention Article 87 (1) (e) accepts that all States enjoy the freedom of fishing on the high seas subject to the requirements codified in section 2 of Part VII. But what if the deployment of pipes would, in one way or the other, spatially affect the deep seabed? While it is difficult to imagine how there could be a link between ocean upwelling pipes and the regime of the Area in practice,[96] the legal issues associated with the question will be addressed briefly for the sake of completeness. Given that LOS Convention Article 153 (1) states that: "[a]ctivities in the Area shall be organized, carried out and controlled by the Authority on behalf of mankind as a whole [...]", the key question would be whether the deployment of ocean upwelling pipes could be regarded as an "activity in the Area". LOS Convention Article 1 (1) defines "activities in the Area" as "all activities of exploration for, and exploitation of, the resources of the Area" and thus incorporates the meaning of "resources" under LOS Convention Article 133 (a) defined as "all solid, liquid or gaseous mineral resources". The use of ocean upwelling pipes is not connected with the exploration for or exploitation of minerals resources on

---

[94] For an in-depth discussion of this submission, see Alexander Proelss, Genetic Resources under UNCLOS and the CBD, in: *German Yearbook of International Law*, Vol. 51, 2008, 417-446, at 430 et seq. A different approach is taken by Alex G. Oude Elferink, The Regime of the Area: Delineating the Scope of Application of the Common Heritage Principle and Freedom of the High Seas, in: *International Journal of Marine and Coastal Law*, Vol. 22, 2007, 143-176.

[95] See also LOS Convention Article 256.

[96] Ibid., Article 1(1) defines "Area" as "the seabed and ocean floor and subsoil thereof, beyond the limits of national jurisdiction".

the deep seabed. Therefore, the deployment and use of these objects cannot be regarded as an "activity in the Area" and remains beyond the limited competence of the International Seabed Authority (ISA).[497]

A different conclusion could only be drawn where ocean upwelling pipes were deployed and used for the purpose of MSR in the Area. In this situation, LOS Convention Article 143 (1) provides that: "[m]arine scientific research in the Area shall be carried out exclusively for peaceful purposes and for the benefit of mankind as a whole, in accordance with Part XIII." However, whether or not this provision can be relied upon independent of the undertaking of activities in the Area is far from clear.[498] It has been argued that it cannot be left unconsidered that LOS Convention Article 143 (1) refers to Article 140 which, as already stated, only applies to "activities in the Area"[499]. Similarly, LOS Convention Article 137 (2) which contains the "mankind as a whole" formula only applies to the "rights in the resources of the Area". Against this background, it is submitted that LOS Convention Article 143 (1) does not give a definite answer as to whether

---

[497] For an overview on the competences of the ISA, see Michael Wood, International Seabed Authority (ISA), in: Wolfrum (ed.), *Max Planck Encyclopedia of Public International Law*, Vol. VI, at 146 and Alexander Proelss, The Role of the Authority in Ocean Governance, in: David D. Caron and Harry Scheiber (eds.), *Institutions and Regions in Ocean Governance* (Martinus Nijhoff Publishers, 2012), in press. See also Tullio Treves, Principles and Objectives of the Legal Regime Governing Areas beyond National Jurisdiction, in: Erik J. Molenaar and Alex G. Oude Elferink (eds.), *The International Legal Regime of Areas beyond National Jurisdiction: Current and Future Developments* (Martinus Nijhoff Publishers, 2010), 7-25, at 16-20.

[498] In the affirmative, see Nele Matz, Marine Biological Resources: Some Reflections on Concepts for the Protection and Sustainable Use of Biological Resources in the Deep Sea, in: *Non-State Actors and International Law*, Vol. 2, 2002, 279-300, at 294.

[499] See Alexander Proelss, supra. note 494, at 425. This issue is further analysed see infra. Chapter III, C (3) (c).

◆Ⅲ. Implementing the MSR Regime in the Marine High-Tech Era◆

MSR *in the Area is subject to the comprehensive jurisdiction of the Authority.* ⁵⁰⁰

With regard to entitlement to exercise jurisdiction over these objects, ocean pipes are equipment in terms of LOS Convention Article 258, if their deployment is conducted for the purpose of MSR. As demonstrated above, according to Article 262 such objects need to bear identification markings indicating the State of registry or the international organization to which they belong. Consequently, if the pipes are located on the high seas, "they are subject to the exclusive jurisdiction of the State of registry in the same way as ships on the high seas are subject to the exclusive jurisdiction of the flag State"⁵⁰¹. For all other purposes, ocean pipes are to be qualified as "devices" in terms of LOS Convention Articles 194 (3) (d) and 209 (2). The wording "vessels, installations, structures and other devices,"⁵⁰² in Article 209 (2) indicates that "device" is a generic concept, which encompasses all categories of objects used in the marine environment. ⁵⁰³ Furthermore, based on the as-

---

⑤⁰⁰ The doubts of the authors are shared by Margaret F. Hayes, Charismatic Microfauna: Marine Genetic Resources and the Law of the Sea, in: Myron H. Nordquist, Ronan Long, Tomas H. Heidar and John N. Moore (eds.), *Law, Science & Ocean Management* (Martinus Nijhoff Publishers, 2007), 683-700, at 690-692; Gwénaëlle Le Gurun, EIA and the International Seabed Authority, in: Kees Bastmeijer and Timo Koivurova (eds.), *Theory and Practice of Transboundary Environmental Impact Assessment* (Brill, 2008), 221-263, at 260; and Ikechi Mgbeoji, (Under)Mining the Seabed? Between the International Seabed Authority Mining Code and Sustainable Bioprospecting of Hydrothermal Vent Ecosystems in the Seabed Area: Taking Precaution Seriously, in: *Ocean Yearbook*, Vol. 18, 2004, 413-452, at 446.

⑤⁰¹ See A. H. A. Soons, supra. note 116, at 234. The same view is also taken by Florian H. Th. Wegelein, supra. note 219, at 150.

⑤⁰² Italics added.

⑤⁰³ See Edward D. Brown, The Significance of a Possible EC EEZ for the Law Relating to Artificial Islands, Installations, and Structures, and to Cables and Pipelines, in the Exclusive Economic Zone, in: *Ocean Development and International Law*, Vol. 23, 1992, 115-144, at 123 and William T. Burke, Highly Migratory Species in the New Law of the Sea, in: *Ocean Development and International Law*, Vol. 14, 1984, 273-314, at 301 and 313.

sumption that all devices operate under the authority of an individual State or international organization, the entitlement for the exercise of jurisdiction over such objects is unambiguously assigned.

Independent of the individual objective pursued with the deployment of ocean upwelling pipes, the freedom to utilize these objects ought to be exercised with due regard for the interests of other States in their exercise of their rights under the LOS Convention.[84] This limitation is particularly important in light of the ever-increasing degree of ocean uses and the consequent emergence of user conflicts. Fishing, aquaculture, energy production, maritime transportation, hydrocarbon activities, MSR, tourism, etc., to name just a few, cannot be carried out without restrictions on marine areas used for ocean upwelling. While one might argue that the regime of the EEZ involves a shift of emphasis in favor of the legal position of the coastal State if and to the extent to which the sovereign rights and jurisdiction of the coastal State are affected and activated by the occurrence of user conflicts,[85] this shift only applies to the deployment of ocean pipes for MSR. For the other purposes for which ocean pipes may be deployed, and with regard to activities on the high seas, LOS Convention Articles 59 and 87 (2) do not provide clear guidelines for the resolution of user conflicts.

At first sight, one might consider that marine spatial planning (MSP) could be a helpful tool to avoid potential conflicts between the deployment of ocean upwelling pipes and shipping routes or other maritime activities. The expanding degree of anthropogenic ocean uses have prompted planners to increasingly regard the oceans as a source of space. According to a definition developed by the United Nations Educational, Scientific and Cultural Organization (UNESCO), MSP constitutes: "a public process of analyzing and allocating the spatial and temporal distribution of human activities in marine areas to achieve ecological, economic and social objectives that have

---

[84] LOS Convention Articles 56 (2), 59, and 87 (2).
[85] See Alexander Proelss, supra. note 83, at 92 et seq.

been specified through a political process."⁵⁰⁶ Of course any MSP activity has to comply with the requirements of the LOS Convention. For example, as regards the EEZ, since a coastal State enjoys a functionally limited sovereign rights and jurisdiction, its domestic law cannot be applied in an undifferentiated manner to that zone.⁵⁰⁷ In contrast, on the high seas, due to the freedom of high seas and in light of the fact that no State is entitled to claim sovereignty over any part of that area pursuant to LOS Convention Article 89, no single State is entitled to unilaterally claim jurisdiction for marine spatial planning.⁵⁰⁸ It is, therefore, necessary that a more convincing mechanism of conflict avoidance be developed. Similar to the case of autonomous underwater vehicles (AUVs),⁵⁰⁹ the most realistic way for the avoidance of user conflicts is the development of non-binding deployment guidelines. Given that the deployment and use of ocean upwelling pipes is related to MSR or constitutes MSR, the Advisory Board of Experts on the Law of the Sea (ABE-LOS) of the UNESCO's Intergovernmental Oceanographic Commission (IOC) should constitute a competent forum for the drafting of these guidelines.

---

⁵⁰⁶ See www.unesco-ioc-marinesp.be (last visited on 15 April 2013).

⁵⁰⁷ See Frank Maes, The International Legal Framework for Marine Spatial Planning, in: *Marine Policy*, Vol. 32, 2008, 797-810, at 799. Note that the fact that Article 56(1), LOS Convention, supra note 15, does not expressly assign to the coastal State a sovereign right or jurisdiction to undertake planning activities does not mean that MSP in the EEZ would be unlawful. Quite the contrary, it has been demonstrated that the LOS Convention contains individual elements of spatial planning and that a general presumption exists in favour of the legality of MSP in the EEZ if and to the extent to which the planning activities concerned are directly linked to the exercise of the sovereign rights and jurisdiction expressly assigned to the coastal State. See Alexander Proelss, supra. note 83, at 110.

⁵⁰⁸ See Frank Maes, ibid., at 799.

⁵⁰⁹ "Draft Guidelines for the Implementation of Resolution XX-6 of the IOC Assembly Regarding the Deployment of Floats in the High Seas within the Framework of the Argo Program." reprinted at IOC/ABE-LOS Ⅷ/3 of 13 June 2008, Eighth Meeting of the Advisory Body of Experts on the Law of the Sea (IOC/ABE-LOS Ⅷ), at 16-17.

(d) Removal Requirements

Ocean upwelling pipes are, in most cases, free floating. This means that the pipes may drift with the surrounding water as a result of currents and waves. Depending on the specific purpose pursued with these objects, their entry into a foreign State's EEZ or territorial sea could affect that State's sovereign rights or jurisdiction. In particular, issues may arise with regard to the recovery of pipes that have floated into the area under another State's jurisdiction. For these reasons, avoidance of unauthorized entrance of upwelling pipes into an area under another State's jurisdiction should be a priority for the placing State.

Reference can be made to the Draft Convention on the Legal Status of Ocean Data Acquisition Systems, Aids and Devices.[510] While the functions of Ocean Data Acquisition Systems (ODAS)[511] are different than those of ocean upwelling pipes, the situation is comparable as both categories of objects may float into other States' areas of jurisdiction. Conditions on the recovery and return of ODAS were explicitly set forth in the Draft Convention with the Convention obliging a State to inform the State of registry about ODAS found under its jurisdiction and to either return the ODAS or permit the owner or operator to recover it.[512] However, ODAS that have entered the internal or territorial waters of a State need not be returned. Prima facie, the parallel challenges that exist with regard to the handling of these objects

---

[510] The text of the 1993 Ocean Data Acquisition Systems, Aids and Devices (ODAS) Draft Convention is available at: unesdoc.unesco.org/images/0009/000979/097992eb.pdf (last visited on 15 April 2013). The goal of the ODA Draft Convention was to comprehensively regulate the legal status of ODAS. See Katharina Bork, supra. note 450, at 316.

[511] Article 1 states that: ODAS means a structure, platform, installation, buoy or other device, not being a ship, together with its appurtenant equipment, deployed at sea for non-military purposes, for peaceful purposes, in non-military service, essentially for the purpose of collecting, storing or transmitting samples or data relating to the marine environment or the atmosphere or the uses thereof.

[512] For details, see Katharina Bork, supra. note 450.

◇ Ⅲ. Implementing the MSR Regime in the Marine High-Tech Era ◇

suggest a call for the adoption of similar regulations vis-à-vis ocean upwelling pipes. Having said that, it should not be forgotten that the ODAS Draft Convention has remained virtually untouched since 1993 and no indications exist that this "treaty ruin" will ever enter into force.[513] Thus, the situation remains unclear though guidelines on the recovery and return of pipes could incorporate the ideas in the ODAS Draft Convention.

Under Part Ⅻ, section 1 of the LOS Convention, States are under an obligation to protect and preserve the marine environment, and are to take all measures to prevent, reduce and control pollution of the marine environment from any source.[514] Moreover, States are to take all measures necessary to ensure that activities under their jurisdiction or control are conducted in a way as to not cause damage by pollution to other States and their environment.[515] In light of these requirements, the removal of ocean pipes is imperative when their life span of around two weeks expires. Given that the pipes are introduced into the marine environment "for a purpose other than the mere disposal thereof", their deployment cannot be regarded as "dumping" in terms of LOS Convention Article 209 in conjunction with Article 1 (1).[516] Nevertheless, LOS Convention Articles 60 (3), 248 (d) and 249 (1) (g) express the general notion that manmade objects intentionally introduced into the marine environment have to be removed once the objective pursued with them has been achieved or they have been abandoned due to expiry of their life span. Moreover, the abandonment of disused or damaged pipes is comparable to plastic garbage[517] as pollution of the marine environment in terms of LOS Convention Article 1 (1).

---

[513] See Katharina Bork, supra. note 468, at 190-191.
[514] See LOS Convention Articles 192 and 194 (1).
[515] Ibid., Article 194 (2).
[516] Ibid., Article 1 (1) No. 5 (b) (ii).
[517] See Annex V, Regulation 3 of the 1973 International Convention for the Prohibition of Pollution from Ships as Modified by the Protocol of 1978 Relating Thereto (MARPOL), 1340 U.N.T.S. 184.

If disused or damaged pipes are located within the deploying State's marine areas, that State is free to remove them. However, if the pipes are located in an area under the jurisdiction of a third State, be it with its authorization or not, the question is who is entitled, or rather obliged, to collect and remove them. Unlike most other installations or structures, ocean pipes are as easy to remove as to deploy, if their location can be ascertained. In light of the sovereignty of the coastal State over the territorial sea, the removal of pipes in that zone falls within the jurisdiction of the coastal State. ⑲ Concerning the EEZ, if and to the extent to which the deployment of ocean pipes is seen as MSR, LOS Convention Article 56 (1) (b) (ii) referring to Articles 248 (d) and 249 (1) (g), places on the research State the obligation to provide the coastal State with information about the expected date of the equipment's removal in order to obtain its consent and to remove scientific research installations and equipment once the research is completed. In contrast, if ocean pipes are used in the context of managing and exploiting the living and non-living resources in the zone, the sovereign rights of the coastal State also includes the decision of when and how to remove the pipes. Finally, if the pipes are used for the purpose of CDR, the fact that such a deployment is neither covered by the sovereign rights and jurisdiction of the coastal State under Article 56 (1), nor by the freedom of communications of third States in terms of Article 58 (1), this militates in favor of accepting that the deploying State is responsible for their removal. It is submitted that the same is true in respect of the high seas.

3. Conclusion

When the LOS Convention was being negotiated, ocean upwelling was not foreseen. While the LOS Convention was, according to its Preamble, concluded in a spirit to "settle [...] all issues relating to the law of the sea"

---

⑲ See Katharina Bork, supra. note 450, at 314. The situation is analogous to that of floats and gliders.

and thus to provide a legal framework applicable also to new developments, there is a clear need to substantiate the general requirements contained in the Convention by way of establishing specific sub-regimes. These need not take the form of binding amendments or a new treaty. The forgoing analysis has advocated a way forward to deal with ocean upwelling through the development of guidelines that respects the structures established by the LOS Convention.

B. Voluntary Observing Ships

1. Factual Background

Ship-based observation, a significant part of synoptic monitoring of the marine environment, has been at the heart of MSR from the start. In the beginning, the navigational safety was the focal point of weather observation at sea. As early as 1853, Matthew Fontaine Maury of the US Navy proposed an international conference to coordinate the establishment of uniform observation systems at sea. This led to the establishment of the International Meteorological Organization, the predecessor of the World Meteorological Organization (WMO). [19] During the past decades, with the further development of marine science and technology, combined with the threat of global warming and other environmental disasters, the requirements of large scale and real-time observations expanded. [20] A well-organized system for performing observations at sea is generally considered necessary nowadays. The Joint Technical Commission for Oceanography and Marine Meteorology (JCOMM), an expert intergovernmental organization

---

[19] For further details, see http://www.bom.gov.au/jcomm/vos/vos.html (last visited on 15 April 2013).

[20] For more information, see http://www.bom.gov.au/jcomm/vos/documents/vos_brochure.pdf (last visited on 15 April 2013).

co-organized by the WMO and the Intergovernmental Oceanographic Commission (IOC), consolidates and coordinates the observations, the data management, and the service system of oceanography and marine meteorology.[51] JCOMM is composed of the Observations Programme Area (OPA), the Data Management Programme Area, and the Services Programme Area, as well as two Cross Cutting Task Teams on Satellite Data Requirements and Capacity Building.[52] The OPA is primarily responsible for developing, coordinating, and maintaining the multi-mode marine observations by means of global cooperation. The Ship Observation Team (SOT) is a subdivision of the OPA and its main component is the Voluntary Observing Ships (VOS) Scheme.[53]

(a) How the VOS Scheme Works

The data collected by VOS are used for preparing forecasts and warnings to help ships to avoid severe weather conditions, to monitor the state of the oceans for climatological databases (which serves many purposes), and to build long term records to monitor changes in the earth's climate.[54] These data pertain to physical properties of the atmosphere above the sea (temperature, dew point, cloud, weather, visibility, and pressure) and the surface of the sea (temperature, waves, currents, and ice).[55]

The members of the WMO recruit all the ships of the VOS scheme. The responsible representative for recruitment is the Port Meteorological

---

[51] For information on the work of JCOMM, see http://www.jcomm.info/ (last visited on 15 April 2013).

[52] For details, see http://www.jcomm.info/index.php?option=com_content&task=view&id=89&Itemid=97. (last visited on 15 April 2013).

[53] For details, see http://www.jcomm.info/index.php?option=com_content&task=view&id=21&Itemid=38 (last visited on 15 April 2013).

[54] See J. Ashley Roach, Defining Scientific Research: Marine Data Collection, in: Myron H. Nordquist, Ronan Long, Tomas H. Heidar and John Norton Moore (eds.), *Law, Science & Ocean Management* (Martinus Nijhoff Publishers, 2007), 541-574, at 557.

[55] Ibid.

◇Ⅲ. Implementing the MSR Regime in the Marine High-Tech Era◇

Officer (PMO), who plays a critical role in running the VOS scheme. The PMO's functions also comprise, amongst other things, maintaining accurate ship records, regularly visiting the ships, and providing relevant service regardless of the ships' nationality and country of recruitment.[526] All the members of the VOS fleet, regardless of the ships' nationalities and the State of recruitment, share their observations from the related routes. The vessels offer their observations in return for obtaining forecasting, warning services, and instrumentations. Besides, there are no direct costs for attending vessels. Communication charges for the transmission are exempt.[527] At present, there are about twenty-five countries listed as having a VOS fleet, with approximately 4000 ships of which around 2000 ships are considered to be active on a monthly basis.[528]

(b) Data Management

Traditionally, the VOS fleet has measured and reported on the atmospheric and sea surface conditions, which are needed for meteorological forecasting.[529] There are two ways how VOS data is provided: one method is through observations, which are transmitted in real time, and the other one

---

[526] More introduction and detailed functions of PMO, see, http://www.bom.gov.au/jcomm/vos/pmo.html. (last visited on 15 April 2013).

[527] For general information see http://www.bom.gov.au/jcomm/vos/vos.html (last visited on 15 April 2013) and http://www.bom.gov.au/jcomm/vos/vos.html (last visited on 15 April 2013). For the situation of USA, on the official website, http://www.vos.noaa.gov/us_vos.shtml (last visited on 15 April 2013). It is said that "VOS operates at no cost to the vessel, with communication charges, observing equipment and reporting supplies furnished by the National Weather Service".

[528] http://www.bom.gov.au/jcomm/vos/documents/jcomm4-brief-vos-ancillary-pilot-project.pdf(last visited on 15 April 2013).

[529] http://www.ncdc.noaa.gov./oa/climate/vosclim/vosclim.html (last visited on 15 April 2013).

is through observations that are recorded on paper or electronic logbooks.[530] In relation to the former, real time observations are transmitted to the National Meteorological and Hydrological Services (NMHSs) by using a Geostationary Technology Satellite (GTS). Some NMHSs keep an archive of the data extracted from the GTS.[531] Regarding the latter, marine meteorological observations are recorded onboard in special logbooks, which are provided by national meteorological services. Thereafter, the recruiting country's PMO collects the logbooks, transfers the observations from the logbooks to magnetic media in a standard format, and sends the data approximately once every three months to global collecting centres in Germany and the United Kingdom. These two centres provide data to eight members who are responsible for preparing climatological summaries.[532]

(c) A "New Situation"

The variables observed by VOS originally focused on the air or the sea surface like dry bulb temperature, dew-point temperature, sea surface temperature, air-sea temperature difference, visibility, weather, wind direction and speed, pressure, cloud coverage, and sea state.[533] Later, with scientific developments, an increased awareness showed that the most challenging scientific problems encompass two or more of the environmental sciences. Further development of climate research will require treating the oceans and atmosphere as a thoroughly interacting system,[534] since the atmospheric

---

[530] See, Elizabeth C. Kent, Graeme Ball, etc., The Voluntary Observing Ship Scheme, see, https://abstracts.congrex.com/scripts/jmevent/abstracts/FCXNL-09A02a-1664333-1-cwp4a07_rev1.pdf (last visited on 15 April 2013), at 1. See also the data flow diagram on http://www.bom.gov.au/jcomm/vos/dataflowhtml (last visited on 15 April 2013).

[531] See Elizabeth C. Kent, Graeme Ball, etc., ibid., at 2.

[532] More detailed information about data management see, http://www.bom.gov.au/jcomm/vos/vos.html (last visited on 15 April 2013); and http://gosic.org/gcos/vos-program-overview.htm(last visited on 15 April 2013).

[533] http://www.bom.gov.au/jcomm/vos/vos.html (last visited on 15 April 2013).

[534] See Thomas A. Clingan, supra. note 67, at 435.

circulation cannot be understood in isolation from considering the ocean, and vice versa.[35] Practically, it is far from enough to conduct climate research only by considering the physical state of the atmosphere and the ocean surface. Therefore, a new situation developed, and scientists began to carry out ocean research via VOS by taking water samples and extending the suite of measurements to chemical and biological properties, such as sea surface salinity, concentration of chlorophyll, dissolved oxygen in the seawater and carbon dioxide partial pressure (pCO_2), etc.[36] In this regard, the most prominent activities underway are CO_2 measurements that are now conducted by many marine research institutions. The oceans are the largest sustained sink of anthropogenic carbon dioxide from the atmosphere. Therefore, understanding the involved physical, chemical, and biological processes, the feedback effects, and the future of this sink is critical for reducing uncertainty in regard to climate change. Scientists normally outfit those VOS with automated CO_2 analyzers as well as thermosalinographs (TSGs) to measure the pCO_2 as well as the temperature and salinity in surface water (i.e. CO_2 also in air) in order to determine the carbon exchange between the ocean and the atmosphere.[37]

The way scientists get onto a vessel with their equipment is quite simple: through negotiation with the shipping company and the captain. This means that applications for the coastal State's consent, as required by

---

[35] See Thomas A. Clingan, supra. note 67, at 435.

[36] For research details see Heike Lueger, Arne Körtzinger, et al., CO_2 Fluxes in the Subtropical and Subarctic North Atlantic Based on Measurements from a Volunteer Observing Ship, in: *Journal of Geophysical Research*, Vol. 111, 2006. Andrew J. Watson, etc., Tracking the Variable North Atlantic Sink for Atmospheric CO_2, in: *Science*, Vol. 326, 2009, at 1391-1393. M. Telszewski, et al., Estimating the Monthly pCO_2 Distribution in the North Atlantic Using a Self-Organizing Neural Network, in: *Biogeosciences*, Vol. 6, 2009, at 1405-1421. T. Steinhoff, et al., Estimating Mixed Layer Nitrate in the North Atlantic Ocean, in: *Biogeosciences*, Vol. 7, 2010, at 795-807.

[37] See http://www.pmel.noaa.gov/co2/story/Volunteer+Observing+Ships+%28VOS%29 (last visited on 15 April 2013).

the LOS Convention through the official channel, are usually not filed before the scientists get onto the vessel to take measurements from the ocean.[38] This practice has existed for about two decades, and not much attention has been paid to it until now. If the company or the captain agrees, the scientists are allowed to embark, and they come to an agreement with the chief engineer on the technical aspects of installing the scientific equipment. The measurements would be conducted all the way along the shipping routes. With respect to the shipping company, some advanced clients require the commercial ships to operate in an environmentally friendly way, which means that the vessels with environmental friendly record are being preferred.[39] Even more importantly, many captains and engineers who care about climate change research are willing to promote and facilitate the development of climate research.[40] All these factors make the VOS-based ocean observations more and more common. However, this method of collecting data is different from the meteorological data collection described in the foregoing, as neither the PMO nor the transmission of data to NMHSs is involved. The scientists who deploy the equipment have their own system of transmitting data, either in a delayed mode or in near-real time using satellite systems.[41] If the data reception in the near-real time mode is not continuous, it alerts the scientists that there is a problem with the equipment. Either they fix the problem, or they can inform the chief engineer to resolve it. A sort of relationship based on mutual understanding exists; hence, the shipping company or the captain voluntarily carries some scientists or equipment to conduct measurements in the ocean. They often provide necessary assistance to the scientists or keep an eye on the automatically running equipment. However, the shipping company or the

---

[38] Conversation with Professor Arne Körtzinger from GEOMAR.
[39] Ibid.
[40] Ibid.
[41] Conversation with Tobias Steinhoff who was a doctoral candidate under the supervision of Professor Arne Körtzinger.

captain is to no extent involved in the data collection.[642]

2. Legal Analysis

The marine environment that is important for marine science and the law of the sea is defined by Soons as the area which "is commonly understood as covering the water column, the seabed and subsoil, and the atmosphere immediately above the sea"[643]. As stated above, variables both relevant to the ocean and meteorology are collected by VOS from the water surface, the water beneath the surface, the atmosphere immediately above the sea, as well as the atmosphere not immediately above the sea. The water surface, the water beneath the surface, and the atmosphere immediately above the sea are, according to Soons' definition, components of the marine environment. That is to say, these variables are collected in the marine environment. As long as the marine environment is involved, the situation might be complicated because of the existence of the LOS Convention. Traditionally, the law of the sea deals with activities on and in the oceans that are tied to the environment in which they take place. The LOS Convention sets out principles and norms for regulating the conduct that is relevant to marine and maritime issues. By virtue of the Preamble to the LOS Convention, the goal of this treaty is to provide the legal order for the uses of the seas and oceans. Thus, taking measurements from the oceans by VOS constitutes an ocean use that may affect marine areas within and beyond the limits of national jurisdiction. Consequently, there is a need to consider the legal issues that flow from it on the basis of the LOS Convention.

As has been shown above,[644] the LOS Convention is silent about what constitutes MSR. Meanwhile, within the VOS scheme, the newly emerged

---

[642] Conversation with Tobias Steinhoff who was a doctoral candidate under the supervision of Professor Arne Körtzinger.

[643] See A. H. A. Soons, supra. note 116, at 124. Also see Katharina Bork, supra. note 450, at 303.

[644] See supra. Chapter I, A.

ocean measurements are conducted without subject to the consent regime as stipulated in the LOS Convention Part XIII. On the one hand, the VOS scheme's original purposes are mainly to enable atmospheric weather forecasting and climate change studies, which allow the free conduct of research in the atmosphere without border limitation. On the other hand, scientists would like to undertake all kinds of existing observations for the sake of climate research, including ocean measurements on VOS. All of these facts put ocean measurements on board of VOS in an awkward position. Is it MSR or not? Is it (and, in the affirmative, where) lawful or not? And how can it be protected?

LOS Convention Articles 245 and 246 make it clear that MSR in the territorial sea, EEZ, and on the continental shelf ought to be conducted under coastal States' consent. In the literal sense, ocean measurements on VOS that are taken along the shipping routes, if regarded as MSR, cannot be seriously seen as constituting lawful conduct by virtue of the LOS Convention. The consideration about the feasible consequences becomes applicable since LOS Convention Articles 27 to 32 codify different rules for merchant ships and for governmental ships that are operated for non-commercial purposes. While government-owned vessels operated for non-commercial purposes enjoy immunities from boarding by other States on the high seas or within marine zones of the coastal States,[98] merchant or civilian vessels do not enjoy such immunities to the same extent.[99] The US Coastal Guard declared that while oceanographic vessels designated as such are exempted from most of the inspection laws of the US, merchant vessels are subject to manning requirements under US law.[100] Namely for commercial ships, which constitute the majority of the VOS fleet's composition, the factor of non-immunities is important to bear in mind. In some cases, it is

---

    [98] LOS Convention Article 32.

    [99] LOS Convention Article 27 and 28. See Montserrat Gorina-Ysern, supra. note 192, at 16.

    [100] Excerpted from Montserrat Gorina-Ysern, supra. note 192, at Note 54.

erroneous that the coastal States' domestic implementation of international law is applicable to foreign vessels for breaching the coastal States' regulations. The measures are beyond the purport of LOS Convention Article 253 which requires coastal States to order the suspension or cessation of the MSR activities, including impounding the vessel and its scientific equipment, issuing very large fines, as well as imprisoning the chief scientist or the master of the vessel.[48] This potentially endangers ocean measurements on VOS. Sometimes coast guards get aboard ships to do routine inspections, e. g. for marine pollution, drug smuggling, or other breaches of regulations. However, the coast guards have no idea about the use of voluntary arrangements or the scientific instruments, and none of the crews or the scientists would mention it.[49] It is therefore concluded that a thorough legal protection is needed. Some regulations are supposed to be worked out to ensure that there is a legal basis for the protection of both researching and coastal States.

(a) Legal Classification of Ocean Measurements on VOS

Firstly, it should be noted that the newly emerging practice of taking ocean measurements on VOS is distinct from traditional data collection activities on VOS in terms of legal status for the following reason: with respect to the VOS scheme, during the LOS Convention negotiations of the Third Committee, the Secretary-General of the WMO expressed the concern that some provisions on MSR might have direct consequences on activities conducted by the WMO over the oceans.[50] Specifically, the VOS Scheme was pointed out, which was carried out by the WMO. The chairman of the Third Committee, realizing that adequate marine meteorological data coverage, including that from areas within the EEZ, was indispensable for timely and accurate storm warnings for navigational safety and protecting lives and property in coastal and offshore areas, expressed that:

---

[48] See Montserrat Gorina-Ysern, supra. note 192, at 17-18.
[49] Conversation with Professor Arne Körtzinger.
[50] See UN Doc. A/conf. 62/80. Also see J. Ashley Roach, supra. note 284, at 203.

> The pertinent provisions of the second revision of the text on marine scientific research would not create any difficulties or obstacles hindering adequate meteorological coverage from the ocean areas, including areas within the exclusive economic zone since such operational and research activities have already been recognized as routine activities within the terms of reference of the World Meteorological Organization and are of common interest of all countries with undoubted universal significance.[651]

While it was decided that collecting marine meteorological data is not MSR regulated by LOS Convention Part XIII,[652] the newly emerging practice of ocean measurements on VOS may face a different situation.

In order to discuss the legal classification regarding this practice, the definition of MSR seems to be, again, inevitably implicated, even though its absence from the LOS Convention is so obvious. As stated above,[653] MSR can be understood as any study or experimental work designed to increase human knowledge of the seabed or the subsoil, the water column, or the atmosphere directly above the water. The exact locations for undertaking MSR are difficult to delineate geographically.[654] Warren Wooster explains that:

> [t]he waters continually move and interchange and the animals add to this a motion of their own. The intimate relation of organisms and environment is evident, but the whole system is fluid and mobile, and the only clear boundary is the land itself.[655]

Scientists need maximum access to all parts of the oceans since oceans

---

[651] See United Nations Conference on the Law of the Sea, Official Records, Vol. XIV, at 102-103, 133-134. Excerpted from J. Ashley Roach, supra. note 284, at 203.

[652] Ibid.

[653] See supra. Chapter III, A (2) (a).

[654] See Florian H. Th. Wegelein, supra. note 219, at 19.

[655] See Wooster, Warren S., Scientific Aspects of Maritime Sovereign Claims, in: *Ocean Development and International Law*, Vol. 1, 1973, 13-20, at 19. Excerpted from Florian H. Th. Wegelein, supra. note 219, at 19.

are connected everywhere and have no natural borders.[56] Therefore, the requirements of ocean area access, flexibility of movement, and port access are extremely important.[57]

As analysed in the foregoing, the emerging practice of ocean measurements on VOS has nothing to do with the WMO's framework. Neither the data collection purpose nor the data management is the same as the VOS scheme, except that they all use the VOS fleets. Based on the ordinary meaning of scientific research,[58] VOS measurements and observations that are intended to increase human knowledge about the ocean and atmosphere can fall within the scope of scientific research. Furthermore, in the light of the definition of marine environment,[59] the conduct of measuring the parameters and taking samples from the ocean or from the atmosphere immediately above the sea can be considered as activities in the marine environment. In that case, the ocean measurements and observations on VOS can be deemed as MSR.[60] No matter which category they belong to, they are all classified as MSR because the notion of MSR in principle covers both "pure" and "applied" scientific research.[61]

(b) Ocean Measurements Conducted on VOS are Basic MSR

Pursuant to the already clarified notion of basic MSR,[62] the research should aim at acquiring new knowledge that is beneficial for all humankind

---

[56] See Herman T. Franssen, supra. note 96, at 153.

[57] See A. H. A. Soons, supra. note 116, at 17. The same view also see Herman T. Franssen, supra. note 96, at 153.

[58] See supra. note 449.

[59] See supra. note 450.

[60] Some parameters, like air temperature, clouds, waves, weather and visibility information, which measured by VOS are not in the marine environment. Therefore, the conduct cannot be deemed as marine scientific research. In this book, only the part, which is deemed as marine scientific research, will be analyzed. For details about what other parameters can be achieved by VOS see Elizabeth C. Kent, Graeme Ball, supra. note 530, at 1 (last visited on 15 April 2013).

[61] See supra. Chapter III, A (2) (a).

[62] See supra. Chapter III, A (2) (a).

and at presenting its feature by prompt availability and full publication of the results. In the case of ocean measurements on VOS, marine data measurements and marine sample analysis are effective means to investigate climate change, from which all humankind may benefit. It has become more and more obvious that climate change is going to affect humankind in many ways, most of which will have negative consequences. As the United Nations Secretary General has said: "[I]t is the major, overriding environmental issue of our time."[59] In terms of research purposes, what concerns coastal States the most are natural resource exploration and threats against territorial security.

(i) Issues Unrelated to Natural Resources Exploitation

Marine natural resources include living resources and non-living resources. Among the living resources, whenever an unknown vessel shows up, whether for research or not, the costal State is more alert to fish catches. These catches are threatened by illegal, unreported, and unregulated (IUU) fishing and overfishing. In this regard, it is obvious that the emphasis of the measurements on VOS is the marine data and samples, rather than fish catches.

With respect to non-living resources, they most importantly comprise mineral resources, which are either dissolved in the seawater, resting on the ocean floor, or found underneath of it.[60] The initial period of non-living resource exploration that concerns coastal States the most is the investigation of the oceans floor's geographical and geological conditions in

---

[59] http://www.unep.org/climatechange/Introduction/tabid/233/language/en-US/Default.aspx (last visited on 15 April 2013).

[60] See Roger H. Charlier & Constance C. Charlier, Ocean Non-Living Resources: Historical Respective on Exploitation, Economics and Environmental Impact, in: *International Journal of Environmental Studies* (Gordon and Breach Science Publishers S. A., 1992), at 123-134. It is said that the dissolved minerals include salt, bromine and magnesium, iodine, potassium, brines, and suspended matter. The minerals from the sea floor include sand, gravel, shells, tin, phosphorus, sulphur, and polymetallic nodules. The minerals beneath the ocean floor include hydrocarbons, coal, and others.

◇Ⅲ. Implementing the MSR Regime in the Marine High-Tech Era◇

order to obtain detailed information on offshore resources.[65] However, learning about the geographical and geological information of the ocean floor is a sophisticated research process, let alone that surveys of potential resources need to be carried out for a considerable amount of time and the technologies have already become so advanced that extra precision instruments are required. It can thus be seen that the variables frequently measured on VOS, like sea surface salinity, dissolved oxygen, and seawater pCO_2, are far from sufficient for exploring or exploiting non-living resources. Therefore, from this perspective, the anxiousness of coastal States seems superfluous.

(ii) Not Endangering the Coastal State's National Security

Every State regards the maintenance of its security as a matter of the highest national priority.[66] Therefore, it is hard to imagine that oceanographic research has no potential military application.[67] Whenever an oceanographer measures depth, water temperature, pressure, or salinity anywhere in the ocean, the collected data could possibly be useful to navies.[68] Even worse, military intelligence or subversive activities could be carried out under the guise of, or along with, MSR.[69] It is thus understandable that the most secure actions for a coastal State appear to be measures, such as restricting or preventing the access of information about nearby waters.

As for the marine measurements on VOS, it should be kept in mind

---

[65] See Teruhis Tsujino, Exploration Technologies for the Utilization of Ocean Floor Resources: Contribution to the Investigation for the Delineation of Continental Shelf, in: *Science & Technology Trends Quarterly Review*, No. 24, 2007, at 68.

[66] See Warren S. Wooster, Michael Redfield, Consequences of Regulating Oceanic Research, in Warren S. Wooster (ed.), *Freedom of Oceanic Research: A Study Conducted by the Center for Marine Affairs of the Scripps Institution of Oceanography* (Crane, Russak&Company, Inc., 1973), 219-234, at 220.

[67] See Herman T. Franssen, supra. note 96, at 161.

[68] Ibid.

[69] See A. H. A. Soons, supra. note 116, at 32.

that its chief purpose is for climate change studies, even though some of the variables taken by VOS may seem suspicious. However, it should not be ignored that there is, as mentioned before, a data management system for marine measurements on VOS as well as the VOS scheme. In terms of the marine data collected on VOS, after each research program researchers would make the data openly accessible to the public through various channels, typically one or two years later.[570] This is often done by putting the data online,[571] which is beneficial to the whole scientific community, or through publishing papers that are considered as scientific literature and as an important source of conducting MSR.[572] It is generally regarded that the "publication of the research findings in the open scientific literature is one indication of the distinction between open and proprietary research"[573]. The tendency towards promoting MSR for the benefit of all humankind is overwhelming, and potentially exceptional cases should not be an excuse for holding up the pace of scientific development. Furthermore, as for conducting ocean measurements on VOS, all the requirements and qualifications pursuant to the LOS Convention, especially Part XIII, will be

---

[570] Conversation with Professor Arne Körtzinger.

[571] Examples like http://www.ioccp.org/UW.html and http://www.carboocean.org (last visited on 15 April 2013). The former one offers coordination service for the ocean carbon community. The latter one is for marine carbon sources and sinks assessment.

[572] There are two phases within the marine scientific research: the collection of data on the one hand, and the interpretation of those data on the other. In many cases, data that already exists can be used for the conduct of marine scientific research. The important source of already existing data is the scientific literature. More information sees A. H. A. Soons, supra. note 116, at 16.

[573] See Herman T. Franssen, supra. note 96, at 168. The same view is also shared by A. H. A. Soons, supra. note 116, at 7. It is argued that "[l]ike is the case with fundamental marine scientific research, the results of such research are generally published or made generally available in another way". And, at page 17, the same view is readdressed that "[a] basic characteristic of open scientific research is the fact that the results are made generally available by publishing them".

observed.

(c) Notifying Regime

Under LOS Convention Article 246, it is expressly provided that MSR in the EEZ and on the continental shelf shall be conducted with the coastal State's consent. Nevertheless, it should be borne in mind that the consent regime is only applicable in circumstances where the research is conducted in the EEZ or on the continental shelf of the coastal State. In the territorial sea, however, the coastal State enjoys sovereignty, which shows that the legal status of the territorial sea is almost like those of the internal waters (except for innocent passage). Consequently, access to the territorial sea and internal waters for foreign vessels for the purpose of conducting MSR, is subject to the coastal State's complete authority.[574] The coastal State may enact any regulations, either requiring prior consent or imposing conditions at its own discretion.[575] Thus, it is much more complex to conduct MSR in the territorial sea or internal waters. Furthermore, as previously mentioned,[576] immunities only apply to warships and other government ships in the territorial sea operated for non-commercial purposes. The reason why this consideration should be made is that commercial ships, which constitute VOS in most cases, would normally enter the respective ports of coastal States, and that means those ships do not only go across the territorial sea but into the internal waters as well.

During the second session of UNCLOS III in 1974, an alternative regime was proposed[577] by seventeen countries that was later referred to as "notifying regime".[578] According to the proposal, "[m]arine scientists would

---

[574] See A. H. A. Soons, supra. note 116, at 46.
[575] Ibid.
[576] See supra. Chapter III, B (2).
[577] See supra. Chapter II, C (5) (b).
[578] They are respectively, Austria, Belgium, Bolivia, Botswana, Denmark, Germany (Federal Republic of), Laos, Lesotho, Liberia, Luxembourg, Nepal, Netherlands, Paraguay, Singapore, Uganda, Upper Volta and Zambia.

be free to conduct any research in the coastal area without first obtaining coastal state consent; such research would be subject only to coastal state notification requirements and a number of conditions"[79]. Specifically, it was stated that:

> [...] the coastal State shall: (a) **Be given at least [...] months' advance notification of the proposed research project;** (b) **Be given as soon as possible a detailed description of the research project, including objectives, methods and instrumentation, locations and time schedule, and information on the research institution concerned and on the scientific staff to be employed;** (c) **Be promptly informed of any major changes with regard to the description of the proposed research project;** (d) **Have the right to participate directly or indirectly in the research project;** (e) **Have access to all data and samples obtained in the course of the research project and be provided, at its request, with duplicable data and divisible samples;** (f) **Be given assistance, at its request, in the interpretation of the results of the research project.**[80]

With the case of ocean measurements on VOS, applying notifying regime seems more appropriate for the following reasons: first, ocean measurements on VOS can be regarded that the measurements are conducted exclusively for peaceful purpose and in order to increase scientific knowledge of the marine environment for the benefit of all mankind. Second, these measurements are done essentially in a continuous way, which means one cruise after another on mostly the same routes. Thus, they have to be regarded as a systematic long-term endeavour. Third, if conducting those measurements on the basis of the coastal State's domestic law or the LOS Convention, the

---

[79] See R. Winner, Science, Sovereignty, and the Third Law of the Sea Conference, in: *Ocean Development and International Law*, Vol. 4, 1977, 297-342, at 309.

[80] UN Doc. A/CONF. 62/C. 3/L19, Article 6. See United Nations Conference on the Law of the Sea, Official Records, Vol. III: Documents of the Conference, at 266.

application procedure would be so prolonged and complicated that it would contradict scientific development. Furthermore, once the delay of clearance request interferes with the research cruise, the quality of the observation results would definitely be impaired since the VOS' cruise is usually regular, and the observations are persistent as well. Accordingly, the clearance procedures can probably be omitted in order to save time and increase efficiency. Still, the notification must be given in advance because the coastal State is entitled to be acquainted with the situation. Nevertheless, it should be noted that when the notifying regime first proposed during the UNCLOS Ⅲ, it was meant for the marine areas beyond the territorial sea where "a coastal State enjoys certain rights over resources in accordance with this Convention".[81] Within the territorial sea, the same proposal addressed that "marine scientific research within the territorial sea established in accordance with this Convention may be conducted only with the consent of the coastal State"[82]. However, in the case of ocean measurements on VOS, due to the peculiarities of the ocean measurements on VOS, notifying regime is proposed to be applicable not only in the coastal State's EEZ and on the continental shelf, but also can be implemented in the territorial sea or even the internal waters.

In the meantime, other obligations and regulations applicable to MSR should continue to be applied to the conduct of marine measurements on VOS. In light of LOS Convention Article 248,[83] the coastal State is entitled to be provided with the full description of the VOS' name, the project's nature and objectives, the method and means to be used, the precise geographical areas, the expected date of conducting the research, the deployment of the equipment and its removal, the name of the sponsoring

---

[81] UN Doc. A/CONF. 62/C. 3/L19, Article 6. See United Nations Conference on the Law of the Sea, Official Records, Vol. Ⅲ: Documents of the Conference, at 266.

[82] Ibid., Article 5.

[83] LOS Convention Article 248 states the duty to provide information to the coastal State.

institution and the person in charge of the project, and lastly, the extent to which the coastal State is considered to be able to participate or to be represented in the project. Moreover, LOS Convention Article 249 addresses the duty to comply with certain conditions imposed on the research State, including to ensure the coastal State's right to participate or be represented in the research project, and even more importantly, to provide the coastal State with all data and samples derived from the MSR project, the preliminary reports, and the final results and conclusions after the research is completed. These conditions are essential elements that could achieve a balance between the interests of the coastal States and the international marine scientific community. [84]

Against this background, it is necessary to draw a parallel between the implementation of implied consent as provided in LOS Convention Article 252 and the notifying regime. Given that the consent regime established in LOS Convention Article 246 has greatly changed the scenario as it used to be to conduct MSR in coastal State's EEZ and on the continental shelf, LOS Convention Article 252 introduces an implied consent regime for proposed MSR projects to attenuate the consent regime. [85] In accordance with LOS

---

[84] See Myron H. Nordquist (ed.), *United Nations Convention on the Law of the Sea 1982: A Commentary*, Vol. IV, supra. note 110, at 540.

[85] See Myron H. Nordquist (ed.), *United Nations Convention on the Law of the Sea 1982: A Commentary*, Vol. IV, supra. note 110, at 562.

Convention Article 248,[98] information regarding the MSR project accompanying a request for clearance for the conduct of MSR activities which are intended to undertake in the coastal State's EEZ or on the continental shelf must be provided six months prior to the date of commencement of the research activities. Pursuant to LOS Convention Article 252, States or competent international organizations may proceed with the MSR project after submitting a clearance request to the coastal State authorities six months in advance if within four months they have not been informed that: (a) it has withheld its consent under the provisions of LOS Convention Article 246; or (b) the information given by the State or competent international organization regarding the nature or objectives of the project does not conform to the manifestly evident facts; or (c) it requires supplementary information relevant to conditions and the information provided for under LOS Convention Article 248 and 249; or (d) outstanding obligations exist with respect to a previous marine scientific research project carried out by that State or competent international organization, with regard to conditions established in LOS Convention Article 249. Montserrat Gorina-Ysern once addressed that:

> **the essential elements for the implementation of this article (LOS Convention Article 252) are threefold. First, implied consent does not apply for MSR intended in the territorial sea, or**

---

[98] According to LOS Convention Article 248, States and competent international organizations which intend to undertake marine scientific research in the exclusive economic zone or on the continental shelf of a coastal State shall, not less than six months in advance of the expected starting data of the marine scientific research project, provide that State with a full description of: (a) the nature and objectives of the project; (b) the method and means to be used, including name, tonnage, type and class of vessels and a description of scientific equipment; (c) the precise geographical areas in which the project is to be conducted; (d) the expected data of first appearance and final departure of the research vessels, or deployment of the equipment and its removal, as appropriate; (e) the name of the sponsoring institution, its director, and the person in charge of the project; and (f) the extent to which it is considered that the coastal State should be able to participate or to be represented in the project.

for any oceanographic research of an applied nature, or involving drilling, the use of explosives, installations and structures on the continental shelf, or the introduction of harmful substances into the marine environment of a coastal State. Second, the researching State must have submitted a clearance request to the coastal State authorities at least six months in advance, with all the required information and with no pending obligations from pervious cruises owed to that coastal State. And third, if all these conditions are met and the coastal State does not respond within four months, the researching State may assume that clearance has been granted.

With respect to the state practice regarding the implied consent regime, a certain degree of flexibility appears, especially in the case of United States. In the Caribbean region, research vessels from United States sometimes sail prior to obtaining formal written approval by the Caribbean Island States. [587] While none of the States involved has formally declared they intend to apply, or are applying LOS Convention Article 252, States practice is evolving towards a timid implementation of the implied consent regime among Caribbean Island States. [588] In terms of UK, it requests the United States to "assume clearance unless the [coastal] authorities respond negatively no later than three weeks before the cruise", although the three weeks period differs from the four months provided in LOS Convention Article 252. [589] Canada who has the longest coastline in the world, allows extremely short lead time for the United States clearance request. [590] Similarly, for the Untied States research vessels, in other regions, say, Central America, South America, and South Pacific, the tendency towards a timid

---

[587] See Montserrat Gorina-Ysern, supra. note 192, at 40-41.
[588] Ibid., at 41.
[589] Ibid.
[590] Ibid., at 159

◆Ⅲ. Implementing the MSR Regime in the Marine High-Tech Era◆

implementation of the implied consent regime continues.[91] However, it should be noted that these practices do not amount to a clear-cut implementation of the implied consent regime under LOS Convention Article 252, only bearing some degree of relevance.[92] The lead time of submitting the clearance request is often less than six months which is inconsistent with LOS Convention Article 252. That is to say, the 6∶4 time frame set up by the LOS Convention Article 252 is not always strictly observed.[93] Besides, most of the arrangements made by the States from forgoing mentioned regions are specifically to the United States government due to proximity or some other reasons. Therefore, these state practices can be deemed as a sui generis implementation of implied consent regime.[94]

It can be seen that there is a certain degree of similarity between the implied consent regime on the one hand, and the proposed notifying regime on the other. They all bound for coastal State's marine areas to conduct MSR projects without formally getting consent from the coastal State. They all need to supply the information regarding the MSR project to the coastal State as stipulated in LOS Convention Article 248 and pursuant to LOS Convention Article 249, and have duty to comply with certain conditions. The huge difference lies in, however, the implied consent regime only applies in the coastal State's EEZ or on the continental shelf, while this notifying regime is proposed to apply not only in the EEZ but also in other marine areas where the coastal State enjoys sovereignty, sovereign rights or jurisdiction. Besides, as for the implied consent regime, the researching State or competent international organization must submit the clearance request first, while for the proposed notifying regime, only needs to send out the notification. Moreover, in the respect of 6∶4 time frame as provided for the implied consent regime in LOS Convention Article 252, it is not

---

[91] See Montserrat Gorina-Ysern, supra. note 192, at 47-100.
[92] Ibid., at 74.
[93] Ibid., at 175-176.
[94] Ibid.

strictly observed by the proposed notifying regime either.

With respect to the legal regime governing ocean measurements on VOS, it cannot simply modify the LOS Convention by substituting the consent regime with the notifying regime. Besides, giving the complexities of the LOS Convention amendment procedure,[595] this does not appear to be a promising solution.[596] Pursuant to LOS Convention Article 243, States and competent international organizations shall cooperate, through the conclusion of bilateral and multilateral agreements, to create favorable conditions for the conduct of marine scientific research in the marine environment and to integrate the efforts of scientists in studying the essence of phenomena and processes occurring in the marine environment and the interrelations between them. Meanwhile, in accordance with LOS Convention Article 311 (3),[597] which echoes with 1969 Vienna Convention on the Law of the Treaties Article 41,[598] two or more States Parties may conclude agreements modifying or suspending the operation of provisions of the LOS Convention, applicable solely to the relations between them, provided that such agreements do not relate to a provision derogation which is incompatible with the effective execution of the object and purpose of LOS Convention. Furthermore, such agreement shall not affect the application of the basic principles embodied in the LOS Convention and the provisions of such agreements should not affect the enjoyment of other State Parties of their rights or the performance of their obligations under the LOS Convention. It is therefore suggested that an agreement between two or more parties of the LOS Convention can be concluded declaring that in terms of ocean measurements on VOS, the notifying regime applies instead of the consent regime established by LOS Convention Part XIII and other than in the

---

[595] See LOS Convention Part XVII, Article 312-316.

[596] See Katharina Bork, supra. note 450, at 315.

[597] LOS Convention Article 311 stipulates relation of the LOS Convention to other conventions and international agreements.

[598] See 1969 Vienna Convention on the Law of Treaties, 1155 U.N.T.S. 311.

◆Ⅲ. Implementing the MSR Regime in the Marine High-Tech Era◆

coastal State's EEZ or on the continental shelf, the notifying regime still enjoys its availability. In the meantime, given the peculiarities of ocean measurements on VOS, this proposition would not incompatible with the effective execution of the objects and purposes of this Convention which according to the Preamble of LOS Convention comprise establishing a legal order for the seas and oceans which will facilitate international communication, and will promote the peaceful uses of the seas and oceans, the equitable and efficient utilization of their resources, the conservation of their living resources, and the study, protection and preservation of the marine environment. Neither would it affect the application of the basic principles embodied in the LOS Convention, even though there are some doubts on what the basic principles might be,[599] nor affect the enjoyment of other State Parties of their rights or the performance of their obligations under LOS Convention. Additionally, LOS Convention Article 311 (4) requires State Parties intending to conclude an agreement referred to in Article 311 (3) to notify other State Parties of their intention to conclude the agreement and of the modification or suspension for which it provides. Such notification shall be made through the UN Secretary-General.[600]

All in all, it is proposed that for the future legal regulation of ocean measurements on VOS, it may be more reasonable to implement the notifying regime by concluding an agreement between two or more State Parties of the LOS Convention, while all the other duties, guidelines and criteria remain the same as for the current regime of MSR codified in the LOS Convention. Arguably, although there is not evident information

---

[599] It was suggested that basic principles are, among others, the principle of the common heritage and the principle of peaceful use of the high seas embodied in LOS Convention Article 88. See D. Freestone and A. Oude Elferink, Flexibility and Innovation in the Law of the Sea—Will the LOS Convention Amendment Procedures Ever Be Used? in: A. Oude Elferink (ed.), *Stability and Change in the Law of the Sea: The Role of the LOS Convention* (Leiden, 2005), at 181. Excerpted from Katharina Bork, supra. note 450, at 315.

[600] See LOS Convention Article 319 (2) (c).

available, this notifying regime may already be valid as far as the existing law is concerned, taking into account the existing practice of VOS with regard to which coastal States have usually not insisted on applying the consent regime.

(d) The Jurisdiction Issue

In terms of jurisdiction regarding vessels on sea, the allocation could be determined in reference to the State's functions in the maritime context. Thus, three players should be kept in mind: the flag State, the coastal State, and the port State. Combined with the legal regime provided in the LOS Convention and in view of the different marine areas, the jurisdiction issue should be considered in a comprehensive and reasonable way.

(i) High Seas—Flag State Jurisdiction

Flag State jurisdiction is the oldest form of maritime jurisdiction. Under the flag State concept, the customary use of the flag has evolved as a means of identification and symbol of the State. It denotes the State whose nationality a ship bears, and whose flag it flies as a symbol of its nationality.[601] The flag State has authority over, and responsibility in respect to, all ships upon which it has conferred nationality.[602] According to LOS Convention Article 87, high seas are open to all States. All States enjoy the freedom of high seas as laid down in a non-exhaustive list under LOS Convention Article 87 (1). Navigational freedom and the right of flag States to sail ships on the high seas are therefore enshrined under the LOS Convention. Churchill and Lowe state that:

> the ascription of nationality to ships is one of the most important means by which public order is maintained at sea. As well as indicating what rights a ship enjoys and to what obligations it is subject, the nationality of a vessel indicates

---

[601] See Doris Koenig, Flag of Ships, in: Wolfrum(ed.), *Max Planck Encyclopedia of Public International Law*, Vol. IV, para. 1, at 99.

[602] Ibid.

◇ III. Implementing the MSR Regime in the Marine High-Tech Era ◇

which State is to exercise flag State jurisdiction over the vessel.[603]

According to LOS Convention Article 91, "[e]very State shall fix the conditions for the grant of its nationality to ships, for the registration of ships in its territory, and for the right to fly its flag" and also provides that "[t]here must exist a genuine link between the State and the ship".[604] The extent of the jurisdiction exercised by the flag State is summarily set out in LOS Convention Article 94.[605] It comprises the obligation to effectively exercise jurisdiction and control in administrative, technical, and social matters,[606] including the construction, equipment and seaworthiness of ships,[607] the manning of ships, labour conditions and the training of crews,[608] as well as the use of signals, the maintenance of communications, and the prevention of collisions.[609] Furthermore, by virtue of LOS Convention Article 94 (5),[610]

it becomes thus clear that the standards, which the State is called to uphold, are international rather than domestic and consequently the ample facility to regulate the ship flying one's flag

---

[603] See R. R. Churchill and A. V. Lowe, supra. note 91, at 257.

[604] The Convention, however, does not impart any solution to the basic problem concerning the definition of genuine link nor does it give any guidance on what conditions for the registration would satisfy the "genuine link" requirement. See Rainer Vogel, Flag States and New Registries, in: Alastair Couper, Edgar Gold (eds.), *The Marine Environment and Sustainable Development: Law, Policy and Science*, 1993, 411-433, at 421.

[605] LOS Convention Article 94 stipulates duties of the flag State.

[606] LOS Convention Article 94 (1).

[607] LOS Convention Article 94 (3) (a).

[608] LOS Convention Article 94 (3) (b).

[609] LOS Convention Article 94 (3) (c).

[610] Art. 94 (5) of the LOS Convention provides that "[i]n taking the measures called for in paragraphs 3 and 4 each State is required to conform to generally accepted international regulations, procedures and practices and to take any steps which may be necessary to secure their observance".

is in actual practice severely curtailed by globally agreed rules and regulations. ⑪

Having established the freedom of high seas, it is clear that the order of the high seas was entrusted primarily to the flag State. Pursuant to Article 92, "ships shall sail under the flag of one State only and, save in exceptional cases expressly provided for international treaties or in this Convention, shall be subject to its exclusive jurisdiction on the high seas". The flag State jurisdiction accordingly becomes "the cornerstone on which the public order of the high seas is erected. Consequently, the strength of the overall regulatory system would necessarily rest upon the effectiveness of flag State jurisdiction". ⑫ The "exclusive jurisdiction", however, could be challenged by offences such as slave trafficking⑬ and piracy⑭, which are both crimes under universal jurisdiction. ⑮ Any other intervention on a vessel on the high seas strictly depends upon the primacy of flag State jurisdiction, ⑯ provided that the UN Security Council has not authorized other States to take respective measures.

Also in view of ocean measurements on VOS on the high seas, it is the principle of flag State jurisdiction that is applicable. VOS are protected on the high seas under international law by the flag State whose flag is entitled to fly. A. W. Anderson states that a "vessel which is so registered and

---

⑪ See Maria Gavouneli, *Functional Jurisdiction in the Law of the Sea* (Martinus Nijhoff Publishers, 2007), at 35.

⑫ Ibid., at 162.

⑬ LOS Convention Article 99.

⑭ LOS Convention Article 105.

⑮ See Maria Gavouneli, supra. note 611, at 159. Other than slave trade and piracy, the stateless vessels are assumed to be equated with pirate or slave vessels, which are subject to universal jurisdiction as well. Of the same view, also see A. W. Anderson, Jurisdiction over Stateless Vessels on the High Seas: An Appraisal under Domestic and International Law, in: *Journal of Maritime Law and Commerce*, Vol. 13, No. 3, 1982, 323-342, at 336.

⑯ See Maria Gavouneli, supra. note 611, at 161.

controlled enjoys a large degree of immunity on the high seas from interference by the vessels of other States"[617]. Furthermore, under the LOS Convention, the freedom of the high seas comprises conducting scientific research as well as marine measurements in this area. It is therefore submitted that while VOS passes through the high seas, conducting ocean measurements is under the flag State's exclusive jurisdiction. Concurrently, since VOS is under the freedom of the high seas as well, no other State can exert intervention therein as long as the conduct is exercised with due regard for the interests of other States and also under the conditions laid down by the LOS Convention and other rules of international law.

(ii) Exclusive Economic Zone

In order to further analyse the jurisdiction issue regarding the ocean measurements on VOS in coastal State's EEZ, the EEZ's legal status should be first ascertained in light of international law, specifically the LOS Convention. It is generally agreed that the EEZ constitutes a sui generis zone, neither a part of the high seas nor of the territorial waters.[618] The EEZ's sui generis legal character has three principle elements: (1) the rights and duties that the LOS Convention accords to the coastal State; (2) the rights and duties that the LOS Convention accords to other States; and (3) residual rights or jurisdiction that do not fall within either of the previous categories.[619]

Under LOS Convention Article 56,[620] the coastal State has acquired

---

[617] See A. W. Anderson, supra. note 615, at 335.

[618] The term "sui generis zone" was first used by the Chairman of the Second Committee of UNLOS Ⅲ. See United Nations Conference on the Law of the Sea, Official Records, Vol.Ⅴ (1976), at 153. The terminology has been accepted by the leading textbooks on the field. See E. D. Brown, *The International Law of the Sea*, Vol.Ⅰ (Dartmouth Publishing, 1994), at 218. Also see R. R. Churchill and A. V. Lowe, supra. note 91, at 166. The note excerpted from Alexander Proelss, supra note 83, at 88.

[619] See R. R. Churchill and A. V. Lowe, supra. note 91, at 166.

[620] LOS Convention Article 56 provides rights, jurisdiction and duties of the coastal State in the exclusive economic zone.

"sovereign rights" on the one hand and "jurisdiction" on the other. However, "sovereign rights" and "jurisdiction" are exercised in a functionally limited way.[61] While the sovereign rights are exercised with regard to exploring, exploiting, conserving, and managing natural resources and other economic activities, the jurisdiction is granted over certain activities, namely, establishing and using artificial islands, installation and structures, MSR, and the protecting and preserving the marine environment. Furthermore, while the notion of "sovereign rights" signifies something less than sovereignty since it could only be exercised once an EEZ was proclaimed,[62] "jurisdiction" denotes an even more restricted exercise of competence.[63] Maria Gavouneli explains that:

> [t]he existence of sovereign rights creates a presumption of sovereignty for the coastal State which would supersede a jurisdiction claim by another State whereas claims of jurisdiction operate on the same level of equality and must be resolved through the standard dispute settlement procedures.[64]

The rights and duties of other States are set out in LOS Convention Article 58, which has been described by Gavouneli as "less specific and certainly more comprehensive".[65] They are primarily concerned with those freedoms expressly provided in LOS Convention Article 87, namely the freedom of "navigation and over flight and of the laying of submarine cables and pipelines, and other internationally lawful uses related to these freedom"[66] with the exception of the freedom of fishing, which has become a sovereign right of the coastal State, and other related topics, including the conduct of MSR as well as the right to construct artificial islands and other

---

[61] See Alexander Proelss, supra note 83, at 89.
[62] See Maria Gavouneli, supra. note 611, at 64-65.
[63] Ibid.
[64] Ibid.
[65] See Maria Gavouneli, supra. note 611, at 64-65.
[66] LOS Convention Article 58 (1).

installations. The "other internationally lawful uses", which are demanded to be "compatible with other provisions of this Convention", include, inter alia, "those associated with the operation of ships, aircraft and submarine cables and pipelines". [627] It is notable that these freedoms exercised by other States in the EEZ are subject to measures relating to coastal States' sovereign rights, [628] as well as the general limitation governing all freedoms of the high seas.

This concurrent jurisdiction in the EEZ, which seeks to strike a balance between coastal States and other States, signifies the true nature of the EEZ. Coastal States and other States are endowed with their own varied rights and duties which are significant for maintaining the order of the EEZ.

As stated earlier, [629] conducting ocean measurements on VOS constitutes MSR and is thus theoretically subject to the coastal State's jurisdiction. The consent regime is supposed to apply, meaning the researching State should seek the coastal State's consent six months before the project starts and should fulfil all the duties and requirements pursuant to LOS Convention Part XIII. However, in practice those measurements are always conducted on VOS during the course of passage through the coastal State's EEZ. According to LOS Convention Article 58 (1), vessels have the freedom of navigation therein. Besides, as mentioned before, [630] the VOS projects are fall out of the MSR regime ambit as provided in LOS Convention Part XIII. Therefore, those measurements are often correspondingly conducted within

---

[627] Ibid. Myron H. Nordquist (ed.), *United Nations Convention on the Law of the Sea 1982: A Commentary*, Vol. II (Martinus Nijhoff Publishers, 1989), at 564.

[628] Ibid., at 565. Article 58 (3) of the LOS Convention stipulates that "[i]n exercising their rights and performing their duties under this Convention in the exclusive economic zone, States shall have due regard to the rights and duties of the coastal State and shall comply with the laws and regulations adopted by the coastal State in accordance with the provisions of this Convention and other rules of international law in so far as they are not incompatible with this Part".

[629] See supra. Chapter III, B (2) (a).

[630] See supra. Chapter III, B (2) (a).

the context of freedom of navigation. As previously established,[30] neither the captain, the ship company, nor the scientist would apply for consent from the likely passing by coastal States. In this sense, the coastal States would not know that this newly emerged conduct—ocean measurements on VOS—exists. Thus it is actually the vessel that voluntarily carries the measurement equipment or the scientists who exercises real control over these ocean measurements. The enforcement power of consent regime as stipulated in LOS Convention Article 246 therefore has been actually considerably curtailed. However, one should keep in mind that a certain regime is oftentimes not applied does not render it automatically invalidity. In this case, potential conflicts would emerge: while the VOS who voluntarily carries the measurement equipment or the scientists, in other words, the flag State, exercises the real control over ocean measurements, pursuant to LOS Convention Article 246, the coastal State as a matter of fact should be the right one to exercise jurisdiction.

Following this line of thought, the proposed notifying regime[32] could represent an alternative to solve this potential conflict. By concluding an agreement between two or more State Parties of the LOS Convention, the relevant States could declare that as much as the ocean measurements on VOS should be considered as MSR, considering its necessity of continuity and the characteristic of increasing human knowledge, these measurements can be conducted by giving out prior notification and in the mean time fulfilling the duties and obligations that are applicable to MSR as stated in LOS Convention Part XIII. In this way, the coastal State in fact allocates a certain degree of jurisdiction over MSR which is supposed to be solely exercised by themselves to the flag State by means of omitting to apply for their consent. Meanwhile, since the duties and obligations derived from MSR regime as provided in LOS Convention Part XIII are still fulfilled by the researching State, the interests of coastal State are appropriately taken care

---

[30] See supra. Chapter III, B (1) (c).
[32] See supra. Chapter III, B (2) (c).

III. Implementing the MSR Regime in the Marine High-Tech Era

of as well.

(iii) Territorial Sea—Coastal State Jurisdiction

The concept of coastal State sovereignty over the territorial sea was accepted for the first time at UNCLOS Ⅰ, which was held in Geneva in April 1958.[633] At UNCLOS Ⅲ, the debates dealt more with the limits of the territorial sea than with its legal status.[634] Eventually, under the LOS Convention, the coastal State's sovereignty over the territorial sea was retained. This was also the case with the breadth of the territorial sea, which is calculated as up to a limit of not exceeding twelve nautical miles, even though State practice is still not uniform.[635] Furthermore, the exercise of coastal State sovereignty over its territorial sea is subject to the right of innocent passage, which can be seen as the main restriction, compared with sovereignty over the land territory as imposed by international law. Therefore, it can be asserted that the right of innocent passage has been introduced to dilute coastal State sovereignty in order to facilitate the navigational rights of all other States. Nevertheless, the right of innocent passage cannot be taken as equivalent to the freedom of navigation in the EEZ or on the high seas, neither in content nor in extent. Hence, it confers the right on the coastal State to "take the necessary steps in its territorial sea to prevent passage which is not innocent"[636] and any contravention of the laws and regulations it enacts would expose the ships to potential punishment.[637] As far as the definition of "innocent" is concerned, LOS Convention Article 19 (1) provides that as long as the passage "is not prejudicial to the peace, good order or security of the coastal State" and "take[s] place in conformity with this Convention and with other rules of international law", the passage is deemed to be innocent. Furthermore, Article 19 (2) enumerates twelve

---

[633] See Haijiang Yang, supra. note 227, at 121.
[634] Ibid., at 123.
[635] Ibid., at 124.
[636] See LOS Convention Article 25 (1).
[637] See Haijiang Yang, supra. note 227, at 147.

types of conduct which would be considered "prejudicial" to the peace, good order, or security of the coastal State, thereby rendering the passage non-innocent. [638]

Having clarified the meaning of "innocent" under the LOS Convention regarding the right of innocent passage, the meaning of "passage" is given more weight in analysing the legal issues of ocean measurements on VOS since as commercial vessel, VOS which are bound for any port would often pass through territorial seas of either the coastal State which is intended to be bound for or other coastal States. The LOS Convention defines "passage" in Article 18 which consists of two categories: lateral passage and passage to or from a call at port facility, either inside or outside internal waters. [639]

Lateral passage is passage without entering internal waters, which is the typical and traditional type of innocent passage in territorial seas. Passage to or from a port facility, either inside or outside internal waters, was introduced as a new component of the definition of passage within the scope of the right of innocent passage at the 1930 Hague Conference. [640] The inclusion may have been motivated by "the desire to facilitate international navigation and trade on the one hand, and correspondingly, to contain

---

[638] With respect to the list of activities in Article 19 (2), the US and the former USSR hold the connotation in an agreement that "Article 19 of the Convention of 1982 sets out in paragraph 2 an exhaustive list of activities that would render passage not innocent. A ship passing through the territorial sea that does not engage in any of those activities is in innocent passage". Contrarily it is believed by Haijiang Yang that the LOS Convention Article 19 (2) only purported to exemplify the vague expression of the concept of innocent passage. By no means is the LOS Convention Article 19 (2) designed to exhaust all cases in which passage would be rendered non-innocent. See Haijiang Yang, supra. note 227, at 159-160.

[639] Ibid., at 149.

[640] Ibid., at 150. The conference for the codification of international law which met at the Hague from March 13 to April 12, 1930 was the first international conference specifically called for that purpose. More details see Hunter Miller, The Hague Codification Conference, in: *The American Journal of International Law*, Vol. 24, No. 4, 1930, 674-693.

III. Implementing the MSR Regime in the Marine High-Tech Era

coastal State jurisdiction over such passage operation on the other". [61] Both of these two forms of passage are of great relevance to the analysis of ocean measurements conducted on VOS. While lateral passage is more relevant when the vessel is bound for one coastal State but passing through another coastal State's territorial sea, passage to or from a call at port facility either inside or outside internal waters of the coastal State is an indispensable course for the purpose of calling at a port facility or roadstead. This means that in some cases, a vessel may cover the both two circumstances of territorial sea passage during the course of being bound for a coastal State.

Regarding ocean measurements on VOS, pursuant to LOS Convention Article 19 (2) (j), [62] carrying out survey activities or research would render the passage non-innocent, and foreign ships would fully fall under the coastal State's jurisdiction if a passage is no longer innocent. [63] Moreover, according to LOS Convention Article 25, [64] the coastal State may take necessary steps to prevent non-innocent passage. Since the Convention is silent on what these "necessary steps" are, it is within the coastal State's discretion to decide what steps should be taken. [65] Given that carrying out ocean measurements on VOS would render the passage non-innocent, either in lateral passage or heading to the port facility, VOS is fully subject to the necessary steps the coastal State may take which would contradict the original intention of VOS in terms of legal status as mentioned previously. [66] It is therefore necessary to consider a set of rules that can be used to mitigate the influence of LOS Convention Article 19 (2) (j) and in the meantime, protect the coastal State's interest.

Similarly, it is still proposed here that the notifying regime applies in

---

[61] See Haijiang Yang, supra. note 227, at 150-151.
[62] LOS Convention Article 19 (2) codifies the situations when the passage of a foreign ship shall be considered as non-innocent.
[63] See R. R. Churchill and A. V. Lowe, supra. note 93, at 87.
[64] LOS Convention Article 25 addresses the rights of protection of the coastal State.
[65] See Haijiang Yang, supra. note 227, at 217.
[66] See supra. Chapter III, B (2) (a).

the case of conducting ocean measurements on VOS in the coastal State's territorial sea. However, it is worth mentioning that the notifying regime first proposed at the second session of UNCLOS Ⅲ was not meant to cover MSR conducted in the territorial sea.[60] In this case, the notifying regime is suggested to extend its availability to the territorial sea. As far as methodology is concerned, a simple teleological reduction of LOS Convention Article 19 is not a promising option. Still, same as in the EEZ, an agreement could be concluded according to LOS Convention Article 311 (3) declaring that the proposed notifying regime is also applicable to the territorial sea and the prior notification should be given before the vessels enter into the coastal State's territorial sea. Accordingly, this extended notifying regime would also guarantee to carry out ocean measurements on VOS continuously in the territorial sea on the one hand and protect coastal State's interest on the other.

(ⅳ) Internal Waters and Ports—Coastal State Jurisdiction

Regarding internal waters, the general absence of a right of innocent passage comprises the principal distinction when compared with the territorial sea.[68] Pursuant to LOS Convention Article 8, waters on the landward side of the territorial sea's baselines, except as provided in Part Ⅳ (Archipelagic States), form part of the internal waters. These water normally comprises different kinds of waters, like bays, estuaries and ports, rivers, and lakes.[69] They are deemed to be an integral part of the coastal State's territory and subject to the same legal regime as its land territory.

---

[60] UN Doc. A/CONF. 62/C. 3/L19, Article 6, para. 1. See United Nations Conference on the Law of the Sea, Official Records, Vol.Ⅲ: Documents of the Conference, at 266.

[68] With the exception of the case that, according to the LOS Convention Article 8 (2), where the establishment of a straight baseline in accordance with the method set forth in Article 7 has the effect of enclosing as internal waters area which had not previously been considered as such, a right of innocent passage as provided in the LOS Convention shall exist in those waters.

[69] See R. R. Churchill and A. V. Lowe, supra. note 93, at 60.

◇Ⅲ. Implementing the MSR Regime in the Marine High-Tech Era◇

As soon as vessels enter the coastal State's internal waters, they place themselves within the coastal State's territorial sovereignty. The coastal State is therefore entitled to enforce its laws and regulations against foreign vessels,[650] even though coastal States commonly enforce their laws and regulations only when its interests are endangered.[651]

Port States are always coastal States, although the reverse is not always true.[652] According to Gavouneli,

> [t]he creation of a separate port State jurisdiction is the direct consequence of the expansion of the coastal State's jurisdiction over the exclusive economic zone, especially in view of the enhanced environmental protection provisions included in the LOS Convention.[653]

The issues concerning the right of accessing ports and exercising jurisdiction over foreign ships in ports have always been surrounded by controversies. Nevertheless, in practice, coastal States generally admit foreign ships to their maritime ports and waterways that are open to international trade and navigation.[654] Although no general right of entry exists in customary law,[655] it is granted under treaty laws that are most commonly bilateral treaties or multilateral conventions and statutes at the international level.[656] It is implied in LOS Convention Part X that land-locked States enjoy the right of access to ports. The same can be said for Article 211 (3), which provides that coastal States shall give due attention to requirements that are considered to be conditions for the entry of foreign vessels into their ports or internal waters or for a call at their off shore terminals.[657]

---

[650] See R. R. Churchill and A. V. Lowe, supra. note 93, at 65-66.
[651] Ibid.
[652] See Maria Gavouneli, supra. note 611, at 44.
[653] Ibid.
[654] See Haijiang Yang, supra. note 227, at 48.
[655] See R. R. Churchill and A. V. Lowe, supra. note 93, at 61.
[656] Ibid. See also Haijiang Yang, supra. note 227, at 56-60.
[657] See also ibid. at 59.

Port State jurisdiction is the competence of States to exercise prescriptive and enforcement jurisdiction over foreign vessels within their ports. It also offers an opportunity for verifying whether visiting foreign ships comply with certain types of national or international technical standards and if they have engaged in certain illegal behaviour in the port State's own maritime zones or beyond.[58] With respect to the innovative feature of the port State jurisdiction, Gavouneli explains that:

> [t]he big difference lies in the "voluntary" character of the ship's presence in port. Whereas the principle of innocent passage shields the seagoing vessel from the jurisdiction of the coastal State, port State jurisdiction strengthens compliance with national rules and regulations without any interference with the freedom of navigation since entry into a port constitutes a voluntary submission of the vessel to the jurisdiction of the port State—either and both running concurrently with the original jurisdiction of the flag State.[59]

Similarly, it is also widely recognized that port States commonly do not exercise jurisdiction with regards to affairs that are essentially internal to ships and that do not affect port State interests.[60] Many commentators have argued that this is merely a matter of courtesy and it does not prejudice a port State's entitlement to exercise such jurisdiction.[61]

Returning to ocean measurements on VOS, as these vessels voluntarily enter a port or internal waters of the coastal State, they accordingly put themselves under the jurisdiction of the port State or coastal State. As stated, port and coastal States normally enjoy the discretion to leave matters

---

[58] See Erik Jaap Molenaar, Port State Jurisdiction, in: Wolfrum (ed.), *Max Planck Encyclopedia of Public International Law*, Vol. VIII, para. 1, at 355.

[59] Maria Gavouneli, supra. note 611, at 44.

[60] See Erik Jaap Molenaar, supra. note 658, para. 11, at 357.

[61] Ibid.

relating purely to a vessel's internal affairs to the flag State authorities.[62] In any event, foreign vessels remain subjected to the flag States' jurisdiction concerning affairs that are essentially internal and do not affect the port or coastal State's interests. Still, port and coastal State jurisdiction will be exercised when the activities affect the peace or good order of the ports or internal waters.[63] Ocean measurements conducted within the port area or internal waters of another State without its consent would normally deemed to be activities that affect the peace and good order of the port State or coastal State. Therefore, those activities are not allowed unless they are regulated between the coastal State and the flag State. When vessels are bound for the coastal State's ports and pass through the territorial sea, the port or coastal State's jurisdiction will still be asserted over the measurements to the same extent as in the territorial sea. Nevertheless, as far as ocean measurements on VOS are concerned, for the sake of increasing human knowledge and promoting MSR, it is suggested that the applicability of notifying regime should be also extended to the internal waters, same as in the territorial sea. In this way, it would facilitate the conduct of ocean measurements on VOS as well as safeguard the port State or coastal State's interest.

3. Conclusion

To date, there are few cases where States have agreed on simplified procedures, either bilaterally or regionally, instead of the LOS Convention procedure.[64] The prospect of implementing the notifying regime concerning ocean measurements on VOS is yet unknown. To a great extent, it may

---

[62] See R. R. Churchill and A. V. Lowe, supra. note 93, at 66. See also Haijiang Yang, supra. note 227, at 84.
[63] Ibid.
[64] See A. H. A. Soons, The Legal Regime of Marine Scientific Research: Current Issues, in: Myron H. Nordquist, Ronan Long, Tomas H. Heidar and John Norton Moore (eds.), *Law, Science & Ocean Management* (Martinus Nijhoff Publishers, 2007), 139-163, at 162.

depend on the States' good will and their concern for marine scientific development.

In light of the ever-increasing degree of ocean uses, which specifically points to ocean measurements on VOS, this work intends to review the legal status of this newly emerging conduct by analyzing the fundamental legal regime of MSR under the LOS Convention. More importantly, the work re-examines the implementation of the LOS Convention regarding the conduct of ocean measurements on VOS which constitute MSR. Several potential resolutions are proposed when vessels pass through the EEZ, the territorial sea, and the internal waters where the coastal State enjoys jurisdiction, sovereign rights, and sovereignty respectively. Based on the special peculiarity of ocean measurements on VOS and compared with the consent regime applied under LOS Convention Part XIII, it is suggested here that the notifying regime serves as guidance for practical considerations and makes the clearance procedures convenient. In spite of the simplified clearance procedures, State Parties are still bound by the duties and obligations under LOS Convention Part XIII. The general provisions in LOS Convention Part XIII provide that States and competent international organizations shall promote and facilitate the development and conduct of MSR in accordance with the LOS Convention.[65] In doing so, all States shall promote international cooperation and create favourable conditions conducting MSR.[66] Ocean measurements on VOS that are conducted exclusively for peaceful purposes and for the benefit of all humankind should therefore be promoted. Otherwise, the researching State should, if requested, provide access to the research data, samples, and results to the coastal State. In the meantime, technical assistance through meaningful cooperation should be encouraged as well.

---

[65] See LOS Convention Article 239.
[66] See LOS Convention Articles 242 and 243.

C. Marine Bioprospecting

1. Factual Background

The ocean is the largest habitat on earth and its biodiversity is extraordinary. Prior to 1977, biologists thought that without solar energy to support a food chain, organisms in the deep sea would depend on organic material descending from surface waters. In 1977, scientists made a discovery on the bottom of the pacific ocean that was considered as one of the most exciting scientific events of the 20th century: [667] they found seafloor vents gushing shimmering, warm, mineral-rich fluids into the cold, dark depths. To their surprise, they found that the vents were brimming with extraordinary, unexpected life. [668] The fluids, which are especially rich in polymetallic sulphides, are spewed out at speeds of up to five metres a second [669] and up to twenty metres high. [670] Since they are black in colour and constitute columnar chimney structures, the vents are often referred to as "Black Smokers". [671] Vents are formed where the earth's crustal plates slowly spread apart and magma wells up from below to form mountain ranges known as mid-ocean ridges. [672] This is why these vents are usually found along the crest of the mid-ocean ridge. What astonished the scientists is that these hydrothermal vents are homes to a great diversity of life. Vent communities

---

[667] See Margaret F. Hayes, supra. note 500, at 683.

[668] More information see http://www.divediscover.whoi.edu/ventcd/vent_discovery/ (last visited on 21 May 2013).

[669] See Pedro Ré, Deep-Sea Hydrothermal Vents "Oases of the Abyss", in Jean-Pierre Beurier, Alexander Kiss and Said Mahmoudi (eds.), *New Technologies and Law of the Marine Environment* (Kluwer Law International, 2000), 67-74, at 67.

[670] See Alexander Proelss, supra. note 494, at 418.

[671] Ibid.

[672] See Pedro Ré, supra. note 669, at 68.

are spectacular for their novelty: 93 percent of the approximately 350 species (to date) are new to science.[673] Vents in different parts of the ocean tend to host different communities.[674] The giant tubeworms can only be found along the East Pacific Rise, while shrimps dominate the Mid-Atlantic Ridge.[675] On land, plants use solar energy to turn water and carbon dioxide into carbohydrates, which are energy for plants and the organisms that eat them. However, in the deep ocean where the sunlight cannot reach, how does a vent food web begin? As stated earlier, the fluids gushing from the vents are especially rich in polymetallic sulphides, which are the primary substance needed for a process called chemosynthesis.[676] Proelss emphasized that:

> [o]rganisms living at hydrothermal vent sites use energy from chemical oxidation instead of light (photosynthesis) to produce organic matter from carbon dioxide (CO_2) and mineral nutrients. The organic matter is then consumed by various organisms with the help of sulphide-oxidizing bacteria which live either in symbiosis with the vent fauna or in the surrounding environment.[677]

The discovery of hydrothermal vents has revealed communities of organisms with unique genetic and biochemical properties. Most importantly, due to their ability to survive under extreme temperatures in high pressure and toxic environments, these organisms represent a seemingly limitless catalogue of medical, pharmaceutical and industrial applications that are quite attractive to commercial entities. Similar repositories of genetic and biochemical resources have been discovered in other deep sea environments. In addition to hydrothermal vents found along mid-ocean ridges, other features such as cold seeps, brine pools, cold water

---

[673] See Pedro Ré, supra. note 669, at 68.
[674] Ibid., at 70.
[675] Ibid.
[676] See Alexander Proelss, supra. note 494, at 418.
[677] Ibid.

corals, and methane hydrate beds are other sites of great interest.[678] Scientists and commercial entities alike are intrigued by this biotechnological potential. Some of these resources are already being sampled for scientific research and commercial purposes by State-sponsored scientific research bodies in conjunction with commercial enterprises.[679]

2. What Is Meant by "Bioprospecting"

The term "bioprospecting" refers to a relatively new concept that is commonly referred to as "biodiversity prospecting".[680] Neither the LOS Convention nor the Convention on Biological Diversity (CBD)[681] defined or even used that term.[682] While there is no internationally agreed definition for either term, bioprospecting was first defined in 1993 as "the exploration of biodiversity for commercially valuable genetic resources and biochemical".[683] Later, in a progress report prepared by the secretariat of the CBD, bioprospecting was defined as:

> the exploration of biodiversity for commercially valuable genetic and biochemical resources. It can be defined as the process of gathering information from the biosphere on the molecular composition of genetic resources for the development

---

[678] See Margaret F. Hayes, supra. note 500, at 684.

[679] See Robin Warner, Protecting the Diversity of the Depths: Environmental Regulation of Bioprospecting and Marine Scientific Research beyond National Jurisdiction, in: *Ocean Yearbook*, Vol. 22, 2008, 411-443, at 411-412.

[680] See Andree Kirchner, Bioprospecting, Marine Scientific Research and the Patentability of Genetic Resources, in Norman A. Martínez Gutiérrez (ed.), *Serving the Rule of International Maritime Law* (Routledge, 2010), 119-128, at 119.

[681] Convention on Biological Diversity, 5 June 1992, UNTS 1970, 79.

[682] See Salvatore Arico and Charlotte Salpin, *Bioprospecting of Genetic Resources in the Deep Seabed: Scientific Legal and Policy Aspects* (United Nations University, Institute of Advanced Studies, 2005), at 15.

[683] See W. V. Reid et al. (eds.), *Biodiversity Prospecting: Using Genetic Resources for Sustainable Development* (World Resources Institute, 1993). Excerpted from Andree Kirchner, supra. note 680.

of new commercial products.[84]

Several definitions are provided in domestic laws as well. Under New Zealand's Biodiversity Strategy, bioprospecting is "the search among biological organisms for commercially valuable compounds, substances or genetic material"[85]. Within the context of the European Union, bioprospecting "entails the research for economically valuable genetic and biochemical resources from nature"[86]. South Africa defined bioprospecting as:

> any research on, or development or application of, indigenous biological resources for commercial or industrial exploitation, and includes systematic search, collection or gathering of such resources or making extractions from such resources for purposes of such research, development or application (...).[87]

The Philippines proposed a definition in their Wildlife Resources Conservation and Protection Act as "research, collection and utilization of biological and genetic resources for purposes of applying the knowledge derived there from solely for commercial purpose".[88] According to Fiji, bioprospecting is "any activity undertaken to harvest or exploit biological resources for commercial purpose[including] investigative research and sampling".[89]

---

[84] See Progress Report on the Implementation of the Programmes of Work on the Biological Diversity of Inland Water Ecosystems, Marine and Coastal Biological Diversity, and Forest Biological Diversity—Information on Marine and Coastal Genetic Resources, Including Bioprospecting. Doc. UNEP/COP/5/INF/7, 20 April 2000.

[85] New Zealand Biodiversity Stratege, 2000. See http://www.biodiversity.govt.nz/pdfs/picture/nzbs-whole.pdf. (last visited on 22 May 2013). Excerpted from Salvatore Arico and Charlotte Salpin, supra. note 682, at 15.

[86] Ibid.

[87] Ibid. Act No. 10 of 2004, National Environmental Management: Biodiversity Act, Article 1.

[88] Ibid. Section 5 (a), Wildlife Resources Conservation and Protection Act, Republic Act No. 9147, 19th March 2001.

[89] Ibid. Fiji's draft Sustainable Development Bill: Integrated and Consolidated Environmental and Resource Management Legislation, 1996.

As much as the existence of divergence regarding whether bioprospecting covers the subsequent stages of the search and sampling of resources, including further application and development,[690] the main element of bioprospecting which is common to all definitions seems to be its connection to commercial activities.[691] The United Nations Secretary-General commented in a 2007 report that:

> **While there is no universally agreed definition of bioprospecting, the term is generally understood, among researchers, as the research for biological compounds of actual or potential value to various applications, in particular commercial applications... In recent years, the term "biodiscovery" has been preferred to "bioprospecting" to put greater emphasis on the investigative aspect of the research and less on the idea of future exploitation...[692]**

Commenting on the difference between biodiscovery and bioprospecting, Ronán Long proposed the following distinction: the phase of initial research and gathering of information could be referred to "biodiscovery", while the term "bioprospecting" could cover the subsequent phases of resource collection for further investigative purposes and eventual commercial application.[693] Prospecting, on the other hand, is defined by the International Seabed Authority's Regulation on Prospecting and Exploration for Polymetallic Nodules as "the research for deposits of polymetallic nodules in the Area, including estimation of the composition, sizes and distributions of polymetallic nodule deposits and their economic values,

---

[690] Ibid.

[691] See Andree Kirchner, supra. note 680, at 120.

[692] See Report of the Secretary-General on Oceans and the Law of the Sea, 12 March 2007. UN Doc A/62/66. Excerpted from Andree Kirchner, supra. note 680, at 120.

[693] See Ronán Long, Regulating Marine Biodiscovery in Sea Areas under Coastal State Jurisdiction, in Myron H. Nordquist, Tommy T. B. Koh and John Norton Moore (eds.), *Freedom of Seas, Passage Rights and the* 1982 *Law of the Sea Convention* (Martinus Hijhoff Publishers, 2009), 133-169, at 146.

without any exclusive rights". ⁶⁸⁴ While economic values can be the subject of prospecting under this definition, subsequent stages of commercial exploitation are not mentioned.

It is clear from legal literature that there is no consensus on the precise meaning of bioprospecting or biodiscovery. However, they share a common element, which is that both of the terms involve research for commercial purposes, irrespective of whether they involve any commercialization activities or not. Viewed from a broad sense, possible elements of a bioprospecting definition include: the systematic search, collection, gathering or sampling of biological resources for the purpose of commercial or industrial exploitation; the screening, isolation, and characterization of commercially useful compounds; testing and trials; and further application and development of the isolated compounds for commercial purposes, including large-scale collection, developing mass culture techniques, and conducting trials for approval for commercial sale. ⁶⁸⁵

Most bioprospecting activities focus on terrestrial organisms. Only since recently may one observe a trend of broadening the bioprospecting focus from terrestrial ecosystems to include marine and freshwater ecosystems. ⁶⁸⁶ An early success was the discovery of the new compound spongothymidine in a species of marine sponge that grew in coastal waters of Florida, and the subsequent development of the oncology drug from this compound in the 1950s. ⁶⁸⁷ In the 1970s, research on marine products accelerated and began to appeal to different disciplines, including

---

⁶⁸⁴ Regulation on Prospecting and Exploration for Polymetallic Nodules, Regulation 1.3 (e). See http://www.isa.org.jm/files/documents/EN/Regs/MiningCode.pdf. (last visited on 23 May 2013).

⁶⁸⁵ See Salvatore Arico and Charlotte Salpin, supra. note 682.

⁶⁸⁶ See An Update on Marine Genetic Resources: Scientific Research, Commercial Uses and a Database on Marine Bioprospecting, a report of United Nations Informal Consultative Process on Oceans and the Law of the Sea, Eight Meeting, United Nations, New York, 25-29, June 2007.

⁶⁸⁷ See Ronán Long, supra. note 693, at 137.

◇Ⅲ. Implementing the MSR Regime in the Marine High-Tech Era◇

biochemistry, biology, ecology, organic chemistry, and pharmacology. [698] 67 patents worldwide were issued for novel compounds for the pharmaceutical industry using marine natural products between 1999 and May 2003. [699] Presently, scientists are increasingly interested in marine biodiversity since marine natural products seem to have a promising future in drug discovery. [700] Most of the interesting molecules from deep sea organisms are still being clinically tested, while some of them have led to the development of products already available on the market. [701] Following a literature review, it can be seen that there is huge potential for marine organisms to hold genetic resources that are valuable for medical, pharmaceutical, industrial, and other applications.

Combined with the foregoing bioprospecting definitions, a few general points can be made regarding marine bioprospecting. First, marine bioprospecting refers to the examination of marine genetic or biological resources (e.g. plants, animals, micro-organisms) for features that may be of commercial value. [702] Second, the precise range of activities may be wide-ranging and extend from the initial sampling of organisms in the marine environment to their subsequent investigation in the laboratory. [703] The means of collection may vary as well. [704] Third, the main focus of marine bioprospecting is on the search for valuable genetic and biochemical information. Therefore, it is the information that is important, not the source material itself. [705]

---

[698] See An Update on Marine Genetic Resources: Scientific Research, Commercial Uses and a Database on Marine Bioprospecting, supra. note 696.
[699] See Alexander Proelss, supra. note 494, at 419.
[700] See An Update on Marine Genetic Resources: Scientific Research, Commercial Uses and a Database on Marine Bioprospecting, supra. note 696.
[701] Ibid.
[702] See Ronán Long, supra. note 693, at 143.
[703] Ibid.
[704] Ibid., at 144.
[705] Ibid.

Given that most authorities mentioned the exploration of marine genetic resources as the purpose of undertaking marine bioprospecting, it is also necessary to clarify the concept of "marine genetic resources". Unlike bioprospecting, genetic resources are defined by the CBD as "genetic material of actual or potential value" while "genetic material" denotes "any material of plant, animal, microbial, or other origin containing functional units of heredity"[706]. In a report of the Secretary-General on Oceans and the Law of the Sea, it was suggested that genetic resources include plant seeds, animal gametes, cuttings, or individual organisms, as well as DNA extracted from a plant, animal, or microbe, such as a chromosome or a gene.[707] It follows that marine genetic resources are marine plants, animals and microorganisms, and parts thereof containing functional units of heredity that are of actual or potential value.[708]

3. Legal Assessment

As marine scientific knowledge increases and developments in oceans technology permit greater access to the ocean and the deep seabed, new and more intensive uses of these areas are bringing consequential impacts on the marine environment. The term "marine bioprospecting" is often used for dual-purpose activities: MSR and commercially exploiting marine resources. It is, however, not clear where MSR ends and the commercial application begins. It is even more difficult to differentiate MSR from marine bioprospecting due to the absence of an internationally agreed definition of MSR. It is not long since the issue of marine bioprospecting began to receive major attention within the realm of the law of the sea and international environmental law. The primary sources of international law applicable to marine bioprospecting are the CBD and the LOS Convention.

---

[706] See CBD Article 2.
[707] See Oceans and the Law of the Sea, Report of the Secretary-General, Addendum, UN Doc A/60/63/Add.1, para.6, 15 July 2005.
[708] See Salvatore Arico and Charlotte Salpin, supra. note 682, at 16.

◇ Ⅲ. Implementing the MSR Regime in the Marine High-Tech Era◇

The CBD entered into force on December 29th, 1993 and has, as of today, 193 parties, including the European Union.[709] Under the three principle objectives of the CBD,[710] it addresses access to genetic resources. However, its main focus to date has been on areas within national jurisdiction,[711] except for one circumstance stipulated in Article 4 (b) of the CBD: "[i]n the case of processes and activities, regardless of where their effects occur, carried out under its jurisdiction or control, within the area of its national jurisdiction or beyond the limits of national jurisdiction."[712] The CBD does therefore apply to activities carried out on the deep seabed by State Parties.[713] While the LOS Convention does not specifically address either marine genetic resources or bioprospecting, there are extensive provisions on MSR. Recently, a view was expressed that an implementing agreement to the LOS Convention should be adopted to address access to, and benefit-sharing from, the utilization of marine genetic resources of areas beyond national jurisdiction.[714] Furthermore, while the international legal focus is on marine bioprospecting activities conducted in areas beyond national jurisdiction, it is important to note that hydrothermal vents and all other types of recently discovered marine ecosystems are also found in the EEZ and on the continental shelf, both within or beyond 200 nautical miles limit.[715] This part examines whether marine bioprospecting is appropriately regulated by existing international legal rules (in particular as far as the relationship to the regime of MSR under the LOS Convention is concerned), and it will take a closer look at the pertinent jurisdictional issues as well.

---

[709] See http://www.cbd.int/convention/parties/list/default.shtml.
[710] The three principle objectives are conservation of biological diversity, the sustainable use of its components, and the fair and equitable sharing of benefits.
[711] See Robin Warner, supra. note 679, at 412.
[712] See Alexander Proelss, supra. note 494, at 422.
[713] Ibid.
[714] See UN Doc. A/AC.276.6, para.12, 10 June 2013.
[715] See Margaret F. Hayes, supra. note 500, at 685.

(a) Legal Classification of Marine Bioprospecting

Attempting to establish a clear-cut differentiation between marine bioprospecting and MSR is a complicated issue because neither of the two concepts is expressly defined in international law. Primary research and sampling in the marine environment is often the first step of both marine bioprospecting and MSR. It is a long process to product development and commercialization after the basic information or research results are obtained.⑯ As noted in the 2005 report of the UN Secretary-General, the difference between scientific research and bioprospecting therefore seems to lie in the use of knowledge and results of such activities, rather than in the practical nature of the activities themselves.⑰

The commercial implications of marine bioprospecting have inevitably led the discussion back to the topic of "pure" and "applied" MSR. Some legal experts and commentators hold the view that marine bioprospecting cannot be considered as MSR.⑱ It has been stated that "[m]arine scientific research has [...] to be distinguished from other investigative marine activities with any kind of commercial component, such as prospecting, exploration, or fish stock assessment, which may involve confidentiality or proprietary rights"⑲. The rationale that lies in this proposition is that:

> **Marine scientific research is characterized by openness, dissemination of data, free exchange of privately owned samples among researchers (in some cases), as well as publication and dissemination of research results. Public availability and free**

---

⑯ See Ronán Long, supra. note 693, at 158.

⑰ See the Report of the Secretary-General on Oceans and the Law of the Sea, 15 July 2005. UN Doc A/60/63/Add.1, para. 202.

⑱ Relevant article see Lyle Glowka, supra note 456, and document see Report of the Secretary-General on Oceans and the Law of the Sea, 4 March 2004. UN Doc A/59/62, para. 261.

⑲ Report of Subsidiary Body on Scientific, Technical and Technological Advice of the Parties of the Convention on Biological Diversity, 22 February 2003. Doc. UNEP/CBD/SBSTTA/8/INF/3/Rev.1, para. 39.

◇ Ⅲ. Implementing the MSR Regime in the Marine High-Tech Era ◇

exchange add to the sum of human scientific knowledge on a particular subject; therefore, it can be presumed to benefit humankind.⑳

This proposition works under the consumption that MSR only entails "pure" MSR, which involves collecting and analyzing information, data, and samples aimed at increasing humankind's knowledge. Due to the commercial component of marine bioprospecting, it cannot be qualified as constituting "pure" MSR. The UN Secretary-General on Oceans and the Law of the Sea drew a parallel between bioprospecting and "ordinary" prospecting in order to disqualify marine bioprospecting as MSR:

> Because of its exploitative purpose and profit-making goals, bioprospecting may be compared to prospecting for mineral resources... Regulation 1 (3) (e) defines prospecting as the search for deposits of polymetallic nodules in the international seabed area, including estimation of the composition, sizes and distribution of polymetallic nodule deposits and their economic values, without any exclusive rights. Although the definition applies specifically to mineral resources, in particular polymetallic nodules, a number of principles implied in the definition can be applicable in the case of marine genetic resources. Thus it is understood that "bioprospecting" does not constitute marine scientific research, but is an investigative activity undertaken for the discovery and estimation of the economic value of a resource, prior to its future commercial exploitation.㉑

Having ascertained that there are many difficulties associated with differentiating pure and applied MSR in practice, the 2007 Report of the UN Secretary-General employed a new approach:

---

⑳ See Lyle Glowka, supra. note 456, at 172.

㉑ See the Report of the Secretary-General on Oceans and the Law of the Sea, 4 March 2004. UN Doc. A/59/62, para. 262.

> UNCLOS provides the legal regime for the conduct of marine scientific research, without defining the term. In the absence of a formal definition, it has been suggested that marine scientific research under UNCLOS encompasses both the study of the marine environment and its resources with a view to increasing humankind's knowledge (so-called "pure" or "fundamental" research), and research for the subsequent exploitation of research (so-called "applied" research).[72]

This sheds significant light on how to interpret what kind of activities can be considered as MSR. Applied research, in this way, can be considered to be in the domain of MSR in a more justifiable way. Relying on the notion of "research for the subsequent exploitation of research" also helps to clarify the legal status of marine bioprospecting since in some cases, it may only become apparent at a much later stage that the knowledge, information and useful materials collected during the pervious research project have a commercial application.[73]

As previously clarified,[74] MSR designed to increase human knowledge with respect to marine environment. Marine bioprospecting often engages in the investigation of marine genetic or biochemical resources to increase knowledge of the concerned marine substance. For example, by studying hydrothermal vents, more knowledge has been expended, such as biology and microbiology of hydrothermal vents fauna, chemistry of hydrothermal vents fluid, fundamental knowledge of biological systems and physiological processes of hydrothermal vents sites.[75] Moreover, geological and geochemical research at hydrothermal vents has the potential for better un-

---

[72] See the Report of the Secretary-General on Oceans and the Law of the Sea, 12 March 2007. UN Doc A/62/66, para. 203.
[73] See Ronán Long, supra. note 693, at 158.
[74] See supra. Chapter III, A (2) (a).
[75] See David Kenneth Leary, *International Law and the Genetic Resources of the Deep Sea* (Martinus Nijhoff Publishers, 2007), at 184-185.

derstanding the genesis of ore deposits and improving models for exploring ores on land.[726] Research may also yield new geological knowledge about the formation, structural deformation, and ageing of the earth's volcanic ocean crust and associated sediments.[727] Since conducting marine bioprospecting can significantly increase human knowledge of the marine environment, it can therefore be regarded as MSR.

Also as established above,[728] applied MSR is supposed to be conducted for a specific practical purpose. In the case of marine bioprospecting, it is conducted for the sake of developing a new product of potentially huge commercial relevance. The actual isolation, characterization, and culture of biological samples (predominately microbial samples) extracted from the deep sea occur either in laboratories operated by public research institutions such as universities, or in laboratories funded by commercial interests.[729] It is observed that currently, scientific research on the genetic diversity of the oceans is mostly State-funded and carried out predominantly by developed countries.[730] Current practices show that State funding usually covers the actual collection and initial description of compounds. However, the process is not linear, and it is often difficult to identify the point at which the private sector comes into play or when it has recognized something of interest from the results of academic research that could trigger its longer-term involvement and investment.[731] Generally speaking, where biotechnology research is funded by the public sector, the results are published in the scientific literature. However, when research is privately funded, results are generally kept confidential and are ordinarily not disclosed until patent

---

[726] See David Kenneth Leary, *International Law and the Genetic Resources of the Deep Sea* (Martinus Nijhoff Publishers, 2007), at 184-185.
[727] Ibid.
[728] See supra. Chapter Ⅲ, A (2) (a).
[729] See An Update on Marine Genetic Resources: Scientific Research, Commercial Uses and a Database on Marine Bioprospecting, supra. note 696.
[730] See UN Doc. A/AC.276.6, para.12, 10 June 2013.
[731] Ibid.

applications have been filed.[132] Marine bioprospecting is thus to be qualified as applied MSR. Against the background that MSR under the LOS Convention context covers both pure and applied MSR, if marine bioprospecting is, arguably, categorized as applied MSR, it is therefore subject to the MSR regime provided in the LOS Convention.

However, the term "applied MSR" comprises two kinds of MSR: the first one is research that may be "applied" in a functional sense but would still serve great benefits for all of humanity, such as chemical oceanographic investigations conducted for marine pollution control measures, physical oceanographic investigations carried out for enhancing long-range weather forecasting, and marine biological investigations for managing living marine resources. The other one is resource-oriented research, which is stipulated in LOS Convention Article 246 (5) (a) as "of direct significance for the exploration and exploitation of natural resources, whether living or nonliving". It is therefore still essential to re-examine the connection between functionally applied research, resource-oriented research, and marine bioprospecting in the context of applied MSR.

At first sight, it seems that marine bioprospecting activities qualify as of "direct significance for the exploration and exploitation of natural resources", since they are conducted for the purpose of developing new products by exploring the marine living resources concerned. However, at closer inspection, difference between exploration and exploitation needs to be further emphasized. As forgoing analysed,[133] exploration can be defined as the "collecting of data concerning natural resources with a view of using them economically".[134] Usually exploration is conducted to provide a basis for the decision whether or not to exploit a nature resource.[135] Exploitation not

---

[132] See An Update on Marine Genetic Resources: Scientific Research, Commercial Uses and a Database on Marine Bioprospecting, supra. note 696.
[133] See supra. Chapter Ⅱ, C (5) (d) (i).
[134] See Florian H. Th. Wegelein, supra. note 219, at 85.
[135] Ibid.

for the purpose of discovery, but clearly links to the utilization of the natural resource.[736] For marine bioprospecting, research purpose is not for natural resources consumption or exploitation. Rather, emphases are placed on the search of biodiversity for valuable genetic and biochemical information. The organisms are considered receptacles of their genes.[737] Scientific attention to these genetic resources in the marine environment is focused on microorganisms found in unusual ecosystems, with a view to culturing and using gene expression to resolve the "supply problem".[738] According to Margaret F. Hayes:

> It is important to seize the functional units of heredity to determine whether and how they can be used or stored while waiting for a future commercial use. For this kind of non-consumptive and almost "intangible" activity there is normally no need of large quantities of living resources, as quality and difference are much more significant for laboratory research than quantity and similarity. Unlike the case of fisheries, the added value of the work on genetic material is tremendous and issues of patents and protection of intellectual property are likely to arise.[739]

This shows that it is the information carried by the marine living or non-living resources that is important, not the source material itself. Other than the actual collection of the marine living or non-living resources, the subsequent lab researches, exploring their further applications as well as development of the isolated compounds for commercial purposes actually bear

---

[736] See Florian H. Th. Wegelein, supra. note 219, at 85.

[737] See Tullio Scovazzi, Mining, Protection of the Environment, Scientific Research and Bioprospecting: Some Considerations on the Role of the International Sea-Bed Authority, in: *The International Journal of Marine and Coastal Law*, Vol. 19, 2004, 383-409, at 400.

[738] See Margaret F. Hayes, supra. note 500, at 686.

[739] Ibid.

more importance. Unlike other living marine resource exploitation, the purpose of marine bioprospecting is not to harvest the whole bodies or tangible parts of the exploited resources for consumption, but to capture information.

As for the question whether there is a distinction between exploration and MSR, it hardly can give a positive answer. Soons defines exploration as "data collection activities concerning natural resources, whether living or non-living, conducted specifically in view of the exploitation" which can clearly be confused with applied MSR.⑪ Indeed, in general terms, there is no evident difference between exploration and MSR.⑪ Marine bioprospecting therefore can be considered as exploration related MSR, but can hardly be regarded as exploitation of natural resources. Additionally, " direct significance", as defined in LOS Convention Article 246 (5) (a), must be limited both in substance and in time.⑫ In terms of substance, marine bioprospecting activities may be relevant to exploration or potential exploitation, but it is questionable in terms of time. This "time" necessitates temporal proximity, namely, that the data must be sufficient to allow for exploitation in the foreseeable future with the technology available.⑬ In the case of marine bioprospecting, research and preclinical tests may take more than a decade, not to mention that only one to two percent of preclinical candidates actually reach the market place.⑭ Uncertainty of the research timeline and outcomes often compromise the prospect of marine bioprospecting activities.

Against this background, there is no way to be certain that marine bioprospecting falls outside the scope of "direct significance for the exploration and exploitation of natural resources" as provided in LOS Convention Article

---

⑩ See A. H. A. Soons, supra. note 116, at 125.
⑪ Ibid.
⑫ See supra. note 395 and 396.
⑬ See Florian H. Th. Wegelein, supra. note 219, at 87.
⑭ See Ronán Long, supra. note 693, at 140.

◆ III. Implementing the MSR Regime in the Marine High-Tech Era◆

246 (5) (a) since it does constitute the exploration of marine resources. However, as far as direct exploitation of natural resources is concerned, marine bioprospecting is not related at all.

It is important to note that marine bioprospecting is mainly done for medical purposes, especially the identification of compounds for treating cancer, Alzheimer's disease, asthma, pain, and viral infections.[745] Its success will greatly benefit all the human beings. It is expressed that the greatest benefits of marine bioprospecting would come from the availability of the products that are developed and the contributions of these products to public health, food security and science.[746] Therefore, under the context of applied MSR, marine bioprospecting can also be categorized as the kind of applied MSR which is functionally "applied" but undoubtedly significant to the benefit of all the human beings. Against the widespread opinion that any intent of economic gain would automatically change the nature of a MSR activity under the LOS Convention regime,[747] this proposition may seem easy to be challenged. However, to develop a new product entails very high costs over a prolonged period. In terms of the marine bioprospecting process, collecting the sample or genetic material is only the beginning of the process. The initial sample screening is followed by additional bioassays that may, but most likely will not, identify compounds that have potential as bioproducts.[748] For potential pharmaceuticals, years of preclinical development, often within an academic setting, precede many more years of analysis and testing.[749] It may take up to fifteen years to produce a new drug result.[750] It is thus fair to say that given all of the financial resources

---

[745] See Ronán Long, supra. note 693, at 138.
[746] See Letter dated 8 June 2012 from the Co-Chairs of the Ad Hoc Open-Ended Informal Working Group to the President of the General Assembly, UN Doc. A/67/95, para. 18, 13 June 2012.
[747] See Tullio Scovazzi, supra. note 737, at 402.
[748] See Margaret F. Hayes, supra. note 500, at 686.
[749] Ibid.
[750] See Ronán Long, supra. note 693, at 140.

required, the intent of economic gain behind the product would seem to be justifiable. This is not to suggest that costs can decide upon whether an activity should be considered as MSR or not. In this case, even if marine bioprospecting activities are conducted with no cost, they still ought to be regarded as MSR since the decisive factor is to increase human knowledge of marine environment. The cost argument only intends to justify the economic incentive, if necessary. Given the characteristic of probably benefiting all the human beings, the economic incentive should not prevent marine bioprospecting from being classified as a kind of applied MSR.

One should not ignore that it is not possible to definitively say how long it will take to make the initial discovery since testing old samples with new technology sometimes reveals research potential that was not evident when the sample was first screened.[50] This may cause a considerable blurring effect on determining how long the applied MSR process will last and at what point the MSR activities are terminated. There is no doubt that pre-clinical research conducted within an academic setting can be regarded as applied MSR. However, it is questionable whether further application and development of the isolated compounds for commercial purposes can still be regarded as applied MSR. MSR, whether pure or applied, is conducted to increase human knowledge. As long as these compounds are applied and developed for commercial purposes, the motive is financial gain rather than increasing human knowledge. It is therefore suggested that the preclinical phase of marine bioprospecting, including the search, collection, gathering, or sampling of genetic or biochemical resources, as well as the screening, isolation and characterization of the commercially useful compounds, can be considered as applied MSR. However, the subsequent large scale applications and development of the isolated compounds for commercial purposes fall out of the MSR ambit.

(b) How to Interpret "Natural Resources" under the LOS Convention

---

[50] See Ronán Long, supra. note 693, at 140.

Ⅲ. Implementing the MSR Regime in the Marine High-Tech Era

in the Context of Marine Bioprospecting

Given the fact that marine bioprospecting is applied MSR which bears huge relevance to the exploration of marine natural resources, it is therefore necessary to examine the term "natural resources". Pursuant to the different marine areas covered by the LOS Convention, this term has four possible separate meanings under the LOS Convention.[652]

LOS Convention Article 56[653] stipulates that in the EEZ, the costal State has sovereign rights for the purpose of exploring, exploiting, conserving, and managing the natural resources, living or non-living, of the waters superjacent to the seabed and of the seabed and its subsoil. Since this context includes both living and non-living natural resources, there is no doubt that the macro-organisms (plants, animals, and fungi) or micro-organisms which are normally the objects of marine bioprospecting can be counted as natural resources under the LOS Convention EEZ regime.

Regarding the continental shelf, LOS Convention Article 77 (1) provides that the coastal State exercises sovereign rights over the continental shelf for the purpose of exploring it and exploitating its natural resources. Unlike in the EEZ, however, Article 77 (4) divides natural resources on the continental shelf into two categories: mineral and other non-living resources of the seabed and subsoil on the one hand, and living organisms belonging to sedentary species on the other.[654] The term "sedentary species" refers to organisms that meet one of two criteria: either they are immobile on or under the seabed, or unable to move unless they keep constant physical contact with the seabed or the subsoil. These criteria are applicable only to organisms that are "at the harvestable state" of their life cycle.[655] S. V. Scott

---

[652] See Ronán Long, supra. note 693, at 140.

[653] LOS Convention Article 56 provides rights, jurisdiction and duties of the coastal State in the exclusive economic zone.

[654] See Myron H. Nordquist (ed.), *United Nations Convention on the Law of the Sea 1982: A Commentary*, Vol. Ⅱ, supra. note 627, at 897.

[655] Ibid.

explains the rationale of including sedentary species on the continental shelf regime:

> It would be senseless to give the coastal State sovereign rights over mineral resources such as the sands of the seabed, but not over the coral, sponges and the living organisms which never moved more than a few inches or a few feet on the floor of the sea.[56]

With respect to sedentary species as part of the natural resources of the continental shelf, the ILC noted that:

> At its fifth session, the Commission decided after long discussion to retain the term "natural resources" as distinct from the more limited term "mineral resources". In its previous draft the Commission had only dealt with "mineral resources" and some members proposed adhering to that course. The Commission, however, came to the conclusion that the products of "sedentary" fisheries, in particular, to the extent that they were natural resources permanently attached to the bed of the sea should not be left outside the scope of the regime adopted, and that this aim could be achieved by using the term "nature resources".[57]

Noting that while sedentary species are included among the continental shelf's natural resources, under LOS Convention Article 68, the notion of "natural resources" does not apply to sedentary species [as covered by

---

[56] See S. V. Scott, The Inclusion of Sedentary Fisheries within the Continental Shelf Doctrine, in: *International and Comparative Law Quarterly*, Vol. 41, 1992, 788-807, at 806.

[57] See Report of the International Law Commission on the Work of its Eighth Session (A/3159), Article 68 and Commentary, para. 3, Yearbook of International Law Commission Vol. II, at 297. http://untreaty.un.org/ILC/reports/english/a_cn4_104.pdf (last visited on 26 May 2013).

◆ III. Implementing the MSR Regime in the Marine High-Tech Era◆

Article 77 (4)] in the EEZ.[58]

This analysis is important because in the marine bioprospecting context, some of the species that inhabit hydrothermal vent communities, seep communities, and deep sea sediment, such as nematodes and molluscs, would be qualified as sedentary species, while others such as the microorganisms in hydrothermal plumes would not.[59] For those who are not sedentary species, they are qualified as natural resources under the EEZ regime.

The third category of natural resources mentioned in the LOS Convention is the "living resources of the high seas" under LOS Convention Part VII, section 2.[60] At first glance, these provisions seem to deal with fisheries and marine mammals. However, there is no evidence that the high seas regime would not apply to other living resources such as hydrothermal vent organisms.[61] According to Proelss:

> It would not have been necessary to exclude sedentary species (which are usually not fish in the biological sense) from the regime of the EEZ[cf. Article 56 (3),77(4) UNCLOS (LOS Convention)], if the term "living resources," used in UNCLOS (LOS Convention), would not comprise resources other than fish.[62]

It follows that living resources under the high seas regime comprise a certain portion of the marine genetic or biochemical resources. The high seas regime is therefore applicable to some of the marine genetic or biochemical resources, and accordingly those marine bioprospecting activities that aim at exploring marine genetic or biochemical resources in the high seas are subject to the freedom of the high seas.

---

[58] See Myron H. Nordquist (ed.), *United Nations Convention on the Law of the Sea* 1982: *A Commentary*, Vol. II, supra. note 627, at 898.

[59] See Robin Warner, supra. note 679, at 419.

[60] LOS Convention Part VII provides the legal regime regarding the high seas. Section 2 is about conservation and management of the living resources of the high seas.

[61] See Alexander Proelss, supra. note 494, at 431.

[62] Ibid.

The fourth category of natural resources in the LOS Convention are the resources of the deep seabed beyond the limits of national jurisdiction, simply referred to as the Area.[163] This is also one of the most contentious elements discussed in the current literature. The Area's resources are defined as "solid, liquid or gaseous mineral resources in situ in the Area at or beneath the seabed, including polymetallic nodules".[164] Resources, when recovered from the Area, are referred to as "minerals".[165] In a literal sense, living resources are not mineral resources and do not come within the scope of the resource definition in LOS Convention Part XI.[166] This relatively restrictive definition of resources in the Area has led to much debate on whether the legal regime of the Area provided by the LOS Convention is applicable to resources other than mineral resources. It has been suggested by Robin Warner that "[t]he jurisdiction ambit of Part XI therefore does not currently extend to living resources located in the Area".[167] It was further noted that:

> **If the species emanating from the chemosynthetic processes of the deep seabed are regarded as having independent life, they are more logically associated with marine living resources under the current provision of UNCLOS [LOS Convention] than with the non-living resources governed by the deep seabed regime under Part XI.**[168]

Churchill and Lowe have suggested that the use of non-mineral resources in the Area falls under the freedom of the high seas regime and is

---

[163] See Ronán Long, supra. note 693, at 162.
[164] See LOS Convention Article 133 (a).
[165] See LOS Convention Article 133 (b).
[166] LOS Convention Part XI stipulates legal regime regarding the Area. See Ronán Long, supra. note 693, at 163.
[167] See Robin Warner, supra. note 679, at 420.
[168] Ibid., at 419.

◇ Ⅲ. Implementing the MSR Regime in the Marine High-Tech Era ◇

excluded from the scope of LOS Convention Part Ⅺ.[69] Alex G. Oude Elferink however has taken a different position:

> **Article 133 does not provide an exhaustive definition of the term "resources" for the purpose of Part Ⅺ, but stipulates that one specific category of resources for the purposes of Part Ⅺ will be referred to as "resources". Article 133 does not state that Part Ⅺ is only applicable to mineral resources. In the case the "expressio unius est exclusio alterius" principle clearly would have been applicable. Article 133, as it is worded, does not exclude living resources or other non-mineral resources from the scope of application of Part Ⅺ.**[70]

It has further been pointed out that by referencing LOS Convention Article 136, which provides that the Area and its resources are the common heritage of mankind, "Article 136 makes the common heritage principle not only applicable to the mineral resources of the Area, but also to the Area as such".[71]

As enlightening as the contentious discussions regarding the interpretation of term "resources" under the regime of the Area appear, it can be seen that from a contextual perspective, the jurisdictional ambit of LOS Convention Part Ⅺ does not currently extend to living resources or other non-mineral resources located in the Area, although some of its provisions regulate the impact of deep seabed mining activities on such resources.[72] Regarding the "common heritage of mankind" concept, pursuant to LOS Convention Article 136, not only the resources but the space itself—

---

[69] See R. R. Churchill and A. V. Lowe, supra. note 91, at 239, footnote 49. The same view shared by Myron H. Nordquist (ed.), *United Nations Convention on the Law of the Sea 1982: A Commentary*, Vol.Ⅲ, supra. note 351, at 29.

[70] See Alex G. Oude Elferink, The Regime of the Area: Delineating the Scope of Application of the Common Heritage Principle and Freedom of the High Seas, in: *The International Journal of Marine and Coastal Law*, Vol. 22, 2007, 143-176, at 152.

[71] Ibid., at 150.

[72] See Robin Warner, supra. note 679, at 420.

the Area—is the common heritage of mankind. While still subject to misunderstandings and different interpretations, it seems clear that the concept applies to the mineral resources located at, on or under the deep seabed.[073] While the "common heritage" concept comprises the Area itself which is defined as "the sea-bed and ocean floor and subsoil thereof, beyond the limits of national jurisdiction",[074] it does not mean that living resources or other non-mineral resources are under the domain of this concept since the definition of the Area does not contain any reference to that space's resources.[075] Hence, it must be concluded that the term "resources in the Area" does not include living and other non-mineral resources.

(c) The Jurisdictional Issues under the LOS Convention

Before delving into jurisdictional issues of marine bioprospecting, it is fundamental to remember that, as established before,[076] not only that marine bioprospecting constitutes applied MSR, but also bears the nature of exploring marine natural resources. Since it is impossible to differentiate applied MSR from exploring marine natural resources, the jurisdiction analysis offered following is based on the assumption that marine bioprospecting is considered as applied MSR, in the meantime, the exploring characteristic would also be taken into account.

Therefore, among various arrays of activities involved in marine bioprospecting, only ocean-related or marine resource-related activities will be examined. According to a summary of intersessional workshops aimed at studying issues relating to the conservation and sustainable use of marine biological diversity beyond areas of national jurisdiction, the process of

---

[073] See Nele Matz-Lueck, The Concept of the Common Heritage of Mankind: Its Viability as a Management Tool for Deep-Sea Genetic Resources, in: Erik J. Molenaar, Alex G. Oude Elferink (eds.) *The International Legal Regime of Areas beyond National Jurisdiction: Current and Future Developments* (Martinus Nijhoff Publishers, 2010), 61-75, at 62.

[074] LOS Convention Article 1 (1).

[075] See Alexander Proelss, supra. note 494, at 424.

[076] See supra. Chapter III, C (3) (a).

marine bioprospecting begins with collection at sea, followed by on-board or shore-based lab analysis. In that regard, while basic analysis could take place on board the research vessel, laboratories on land are better equipped for a detailed analysis of the samples. In most cases, samples are collected and preserved on board for future analysis on land, with the exception of live organisms, which require immediate analysis on board the research vessel. [777] Consequently, considering the continuation of marine bioprospecting, some of the lab researches will be included as well.

Whereas much of the debate on this subject focuses on sea areas beyond national jurisdiction, with particular emphasis on the regime that ought to apply to deep ocean hydrothermal vent sites, cold water seeps, deep water corals and seamounts,[778] there has been little discussion of the legal regime that applies to marine bioprospecting in sea areas under coastal State jurisdiction. It should not be ignored that many of the early research programmes focused on shallow tropical water mainly because of easy access. [779] Only more recently has the focus shifted towards exploring the continental shelf and the deep ocean floor, where various forms of life live in extreme environments, due to rapid developments in marine technology and ocean science. [780]

(i) Internal Waters and Territorial Sea

Pursuant to the relevant rules of the LOS Convention, the coastal State enjoys sovereignty over its internal waters and territorial sea. [781] This sovereignty extends not only to the internal waters, territorial sea, and air space, but also to the seabed and subsoil of the marine areas concerned. [782] By its very nature, the coastal State has the competence to prescribe and enforce

---

[777] See UN Doc. A/AC. 276. 6, para. 12, 10 June 2013.
[778] See Ronán Long, supra. note 693, at 134.
[779] Ibid., at 138.
[780] Ibid.
[781] LOS Convention Article 2 (1).
[782] LOS Convention Article 2 (2).

its national regulations regarding exploration, exploitation, and management of living or non-living resources. Considering the exploration nature, marine bioprospecting in coastal State's internal waters and territorial sea is therefore totally subject to coastal State's sovereignty.

Pursuant to the sovereignty, LOS Convention Article 245[83] also gives coastal States the exclusive right to regulate, authorize, and conduct MSR in their territorial sea. Other States wishing to undertake MSR can only do so "with express consent of and under the conditions set forth by the coastal State".[84] As previously mentioned, with respect to marine bioprospecting, two processes fall within the ambit of MSR: searching, collection, gathering or sampling genetic or biochemical resources from the ocean, and screening, isolating, and characterizing commercially useful compounds, which is usually undertaken in research labs. In this sense, undertaking these activities is fully subject to the coastal State's sovereignty, regardless of whether they are conducted in the water column or on the seabed or subsoil.

In this sense, marine bioprospecting conducted in coastal State's internal waters or territorial sea is fully subject to the coastal State's sovereignty, no matter under the consideration of MSR regime or the sovereignty enjoyed by the coastal State deriving from marine natural resources exploration perspective.

(ii) EEZ

According to LOS Convention Article 56 (1) (a), the coastal State enjoys sovereign rights in its EEZ for the purpose of "exploring and exploiting, conserving and managing the natural resources, whether living or non-living, of the waters superjacent to the sea-bed and of the sea-bed and its subsoil". Consequently, marine bioprospecting, which bears the purpose

---

[83] LOS Convention Article 245 provides that: "Coastal States, in the exercise of their sovereignty, have the exclusive right to regulate, authorize and conduct marine scientific research in the territorial sea. Marine scientific research therein shall be conducted only with the express consent of and under the conditions set forth by the coastal State."

[84] Ibid.

of exploring natural resources, should be subject to coastal State's sovereign rights. Meanwhile, pursuant to LOS Convention Article 56 (1) (b) (ii), coastal State also has jurisdiction over MSR. Considering that marine bioprospecting is regarded as applied MSR, coastal State therefore should enjoy jurisdiction over marine bioprospecting as well. Likewise, LOS Convention Article 246 has echoed this jurisdiction by providing the consent regime. In this case, jurisdictional conflict emerges, namely sovereign rights under LOS Convention Article 56 (1) (a) versus jurisdiction under LOS Convention Article 56 (1) (b) (ii). As stated above,[785] there is a difference in quality between sovereign rights and jurisdiction. The existence of sovereign rights creates a presumption of sovereignty, which would supersede a jurisdictional claim.[786] In other words, in the case of marine bioprospecting, the sovereign rights enjoyed by the coastal State with respect to exploration of natural resources take precedence of jurisdiction vested to the coastal State regarding MSR. This assumption has been substantiated by Soons stating that:

> Although the term "marine scientific research", in its ordinary meaning, would cover any scientific investigation of the marine environment, both fundamental and applied, it appears from the provisions of the Draft Convention [LOS Convention] rules that for the purpose of the application of the Draft Convention [LOS Convention] certain applied scientific research activities are excluded from its scope. The Draft Convention [LOS Convention] provides for a separate regime for resources exploration. Resource exploration in all areas under coastal State jurisdiction is subject to the sovereign rights of the coastal State.[787]

Florian H. Th. Wegelein has also expressed the same view that:

---

[785] See supra. Chapter Ⅲ, B (2) (d) (ii).
[786] See Maria Gavouneli, supra. note 611, at 65.
[787] See A. H. A. Soons, supra. note 116, at 125.

On the basis of the exclusive economic zone concept, namely, that it constitutes a zone sui generis within which uncertainties in law need to be settled in the light of the whole balance struck between coastal state rights and the freedom of the high seas, one could take the view that Article 73 of the 1982 LOS Convention is also applicable in respect of research activities... pertain to the living resources of the exclusive economic zone, a distinction between marine scientific research and exploration is difficult to make anyway.[788]

Accordingly, it can be seen that marine bioprospecting conducted in the coastal State's EEZ is subject to the coastal State's sovereign rights based on LOS Convention Article 56 (1) (a). The consent regime stipulated in LOS Convention Article 246 is not applied anymore. Although sovereign rights denote something less than full sovereignty since such sovereign rights could only be exercised once an EEZ was proclaimed,[789] compared with the consent regime, the coastal State enjoys more discretionary power over governing MSR.

Moreover, it needs to be addressed that the EEZ regime in principle comprises the seabed and its subsoil.[790] Churchill and Lowe have confirmed this view:

> Had it not been for a strong desire on the part of many coastal States, now reflected in the provisions of the Law of the Sea Convention, to include within the legal continental shelf those parts of the continental margin extending beyond 200 miles, the

---

[788] See Florian H. Th. Wegelein, supra. note 219, at 186. LOS Convention Article 73 provides a number of enforcement measures in exercise of coastal State's sovereign rights. The language used in LOS Convention Article 73 (1) is parallel to that used in LOS Convention Article 56 (1) (a), whereas MSR is listed in the same Article in subparagraph (1) (b) (ii).

[789] See Maria Gavouneli, supra. note 611, at 64-65.

[790] See Alexander Proelss, supra. note 494, at 430, footnote 55.

legal regime of the continental shelf could have been subsumed within the EEZ.⁽⁷⁹⁾

LOS Convention Article 246 simultaneously addresses the balancing of the coastal States' and other States' substantive rights, both in the EEZ and on the continental shelf, which echoes to the foregoing rationale. Regarding marine bioprospecting, the research object primarily covers marine living resources, together with some sedentary species. Though the sovereign rights entitled by the coastal States in their EEZ comprise exploring, exploiting, conserving, and managing the seabed and subsoil's natural resources, these rights, pursuant to LOS Convention Article 68, do not apply to sedentary species as defined in LOS Convention Article 77 (4).⁽⁷⁹⁾ It is therefore submitted that "[A]rticle 68, read together with [A]rticle 77, retains the rule that sedentary species as defined in [A]rticle 77 come within the regime of the continental shelf".⁽⁷⁹⁾ Following this line of consideration, if marine bioprospecting activities that constitute MSR are conducted for the purpose of researching sedentary species located in the coastal State's EEZ, they are subject to coastal State's sovereign rights which deriving from LOS Convention Part Ⅵ, continental shelf regime, not the EEZ regime as provided in LOS Convention Part Ⅴ.

Additionally, the coastal State's sovereign rights with respect to conserving and managing living resources in EEZ, stipulated by LOS Convention Articles 61-67 mainly refer to fish and marine mammals. Nevertheless, from a marine bioprospecting perspective, a wide range of marine organisms are reservoirs of marine genetic or biochemical resources with actual or potential value, such as bacteria, fungi, micro-and macro-algae, cnidaria (which includes animals such as corals, sea anemones and

---

⑲　See R. R. Churchill and A. V. Lowe, supra. note 91, at 166.

⑳　Sedentary species as provided in LOS Convention Article 77 (4) is analysed in supra. Chapter Ⅲ, C (3) (b).

㉑　See Myron H. Nordquist (ed.), *United Nations Convention on the Law of the Sea 1982: A Commentary*, Vol. Ⅱ, supra. note 627, at 689.

jellyfish), porifera, ascidians, molluscs, worms, fish, and mammals.⑱ Accordingly, in terms of marine living resources conservation, LOS Convention Articles 61-67 could not fulfill the needs for marine bioprospecting since under the marine bioprospecting context, living resources comprise not only fish and marine mammals, but also other organisms. Moreover, LOS Convention Articles 61-67 were provided under the assumption that utilizing marine living resources would be substantial without regulation. However, unlike fishing, marine bioprospecting usually only requires small quantities of sediment, water (in the case of micro-organisms), or individuals for laboratory analysis because it is the heredity units that are targeted.⑲ Therefore, in terms of marine living resources utilization, marine bioprospecting does not pose as much a threat as fishing does. The "utilization of the living resources" regulation under LOS Convention Article 62 does not suit for marine bioprospecting activities. Consequently, marine bioprospecting activities do not fit neatly into the EEZ regime with respect to living resources conservation and management.

(iii) Continental Shelf within or beyond 200 Nautical Miles

Similar to the EEZ, on the continental shelf sovereign rights enjoyed by the coastal State prevail over its jurisdiction. That is to say, as far as marine bioprospecting on the continental shelf is concerned, since bearing the nature of exploring marine natural resources, it is subject to the coastal State's sovereign rights rather than the consent regime as stated in LOS Convention Article 246. Besides, according to LOS Convention Article 77 (2) and (3), coastal State enjoys exclusive rights of exploring natural resources on the continental shelf, namely, if the coastal State does not explore the sedentary species, no one may undertake these activities without the express consent of the coastal State. Furthermore, unlike in the EEZ, the rights do not depend on occupation, effective or notional, or any express proclamation.

---

⑱ See the Report of the Secretary-General on Oceans and the Law of the Sea, 12 March 2007. UN Doc A/62/66, para. 170-178.

⑲ Ibid., para. 190.

It should further notice that the notion of "natural resources" plays a critical role in interpreting jurisdictional issue over marine bioprospecting on the continental shelf, especially when a coastal State has successfully claimed the outer continental shelf. Pursuant to LOS Convention Article 77 (4), natural resources under continental shelf context include non-living resources as well as sedentary species. However, Hayes argued that "different species of scallops are sedentary or not, depending on their exact method of locomotion", and to categorize marine genetic resources as sedentary or non-sedentary is even more of a challenge.[796] Thus, there may be a grey area when it comes to distinguishing between sedentary and non-sedentary resources. Namely, if a certain sedentary species is not necessarily immobile or unable to move except in constant physical contact with the seabed or subsoil, then MSR marine bioprospecting activities conducted on the outer continental shelf may subject to the freedom of high seas.

(iv) Marine Areas beyond National Jurisdiction (ABNJ)

"Areas beyond national jurisdiction" (ABNJ) comprise the water column beyond the coastal State's EEZ (or its territorial sea if a coastal State has not exercised its right to claim an EEZ), and the seabed, ocean floor and subsoil of the submarine areas beyond coastal State's continental shelf, which is referred to as the Area in the LOS Convention Part XI. Altogether, ABNJ cover some 64% of the ocean's surface and over 90% of its volume.[797] LOS Convention Parts VII and XI address legal regimes regarding the high seas and the Area, respectively.

According to LOS Convention Article 257,[798] all States, irrespective of their geographical location, and competent international organizations have

---

[796] See Margaret F. Hayes, supra note 500, at 689.

[797] Ibid.

[798] LOS Convention Article 257 provides that: "All states, irrespective of their geographical location, and competent international organizations have the right, in conformity with this Convention, to conduct marine scientific research in the water column beyond the limits of the exclusive economic zone."

the right to conduct MSR in the water column beyond the limits of the EEZ. Marine bioprospecting can therefore be conducted under the freedom of high seas, in conformity with other provisions of the LOS Convention. However, other than this freedom, the relevance of the high seas regime to marine bioprospecting activities becomes apparent in light of the conservation and management of the living resources. As established before, the notion of living resources under the high seas regime should not be interpreted as restrictive as it appears. It is broad enough to include free-living and symbiotic microorganisms.[799] A certain number of marine genetic or biochemical resources can be regarded as living resources under the high seas regime. In this sense, in accordance with LOS Convention Article 117 and 118,[800] State Parties are obliged to take conservation and management measures regarding marine genetic and biochemical resources in the high seas, both domestically and in cooperation with other State Parties.

The deep seabed that hosts a wide array of biological communities attracts significant scientific attention. Exploration activities related to deep seabed ecosystems were described as "scattered, small scale, independent research activities and programmes ongoing in many universities and research institution in the world".[801] As for the MSR regime in the Area, according to LOS Convention Article 256,[802] States have the right to conduct MSR in the Area as long as it is in conformity with the provisions of LOS

---

[799] See Lyle Glowka, supra note 456, at 168.

[800] LOS Convention Article 117 codifies the duty of States to adopt with respect to their nationals measures for the conservation of the living resources of the high seas. Article 118 is about cooperation of States in the conservation and management of living resources.

[801] See Salvatore Arico and Charlotte Salpin, supra. note 682. Excerpted from Robin Warner, supra. note 679, at 416.

[802] LOS Convention Article 256 provides legal regime regarding marine scientific research in the Area. It stipulates that: "All States, irrespective of their geographical location, and competent international organizations have the right, in conformity with the provisions of Part XI, to conduct marine scientific research in the Area."

◇ Ⅲ. Implementing the MSR Regime in the Marine High-Tech Era◇

Convention Part ⅩⅠ.

Regarding the admissibility of LOS Convention Article 143 in terms of marine bioprospecting, a closer examination seems necessary. While "for the benefit of mankind as a whole" may be questionable because of the promising commercial prospects for marine bioprospectors or pharmaceutical firms involved, it cannot be denied that the results will eventually benefit all human beings. As stated in a document submitted by the SBSTTA of the CBD:

> **A brief search of Patent Office Databases revealed that compounds from deep seabed organisms have been used as basis for potent cancer fighting drugs, commercial skin protection products providing higher resistance to ultraviolet and heat exposure, and for preventing skin inflammation, detoxification agents for snake venom, anti-viral compounds, anti-allergy agents and anti-coagulant agents, as well as industrial applications for reducing viscosity.**[803]

However, as far as interpreting "for the benefit of mankind as a whole" under the LOS Convention is concerned, a different position could be taken as briefly mentioned before.[804] LOS Convention Article 143 indirectly refers to Article 140, which elaborates on the "common heritage of humankind" principle.[805] According to Article 140, activities in the Area are to be carried out for the benefit of humankind as a whole, and the Authority is to ensure equitable sharing of financial and other economic benefits derived from

---

[803] See Status and Trends of, and Threats to, Deep Seabed Genetic Resources beyond National Jurisdiction, and Identification of Technical Options for their Conservation and Sustainable Use, para. 21. Doc. UNEP/CBD/SBSTTA/11/11 of 22 July 2005. Excerpted from Tullio Scovazzi, The Seabed beyond the Limits of National Jurisdiction: General and Institutional Aspects, in: Erik J. Molenaar, Alex G. Oude Elferink (eds.), *The International Legal Regime of Areas beyond National Jurisdiction: Current and Future Developments* (Martinus Nijhoff Publishers, 2010), 43-60, at 52.

[804] See supra. Chapter Ⅱ, C (3) (a).

[805] See Myron H. Nordquist(ed.), *United Nations Convention on the Law of the Sea 1982: A Commentary*, Vol. Ⅵ, (Martinus Nijhoff Publishers, 2002), at 131-132.

activities in the Area. Yet, LOS Convention Article 143 seems to restrict "the benefit of humankind" formula applicable to "activities in the Area". [86] As defined in LOS Convention Article 1 (1) (3), "activities in the Area" means all exploration for, and exploitation of, the resources of the Area. If one incorporates the restrictive meaning of "resources" under LOS Convention Article 133, "activities in the Area" is limited to the exploration for, and exploitation of, the Area's mineral resources, and the same to be true for the "benefit of mankind" formula. [87] Thus, if one follows a more restrictive interpretation of "benefit of mankind" as embodied in LOS Convention Article 140, it would seem questionable whether LOS Convention Article 143 is still applied to marine bioprospecting activities. Moreover, under LOS Convention Article 143 (3), States have the right to carry out MSR in the Area, but are, at the same time, bound to cooperate with other States and the Authority, and to effectively disseminate the research and analysis results when made available through the Authority or other international channels. In this sense, the dissemination of provision does not seem feasible due to the commercial nature of marine bioprospecting activities. Therefore, following this line of reasoning, it is still far from confirmative on whether and to what extent LOS Convention Article 143 would apply to MSR marine bioprospecting. Given that there does not seem to be a sufficient body of State practice, at least one can conclude that reading Article 143 in combination with Article 246 would contradict the unwarranted assumption that there is an absolute freedom to carry out marine bioprospecting in the Area. [88]

Marine bioprospecting in the Area is frequently conducted for both scientific and commercial purposes. [89] Besides, due to the special legal status of the Area, marine bioprospecting in the Area has led to more discussions

---

[86] See Alexander Proelss, supra. note 494, at 424.
[87] Ibid.
[88] See Tullio Scovazzi, supra. note 803, at 58.
[89] See Robin Warner, supra. note 679, at 421.

III. Implementing the MSR Regime in the Marine High-Tech Era

regarding its legal regulation. As far as the governance of marine bioprospecting activities is concerned, two different, albeit interconnected, subjects are relevant: the regime for marine genetic or biochemical resources in the Area on the one hand and the idea of a general regime of the Area on the other hand.⑩ The main divergence lies in whether the principle of common heritage of mankind or the freedom of the high seas should apply to marine genetic and biochemical resources.⑪

As stated, a simple interpretation from a contextual perspective would be, under LOS Convention Part XI, that the notion of "resources" is so restrictive as to only refer to mineral resources. Accordingly, the Area regime does not encompass resources other than mineral resources, such as marine genetic or biochemical resources that oftentimes exist in the form of living resources or sedentary species. It is thus fair to say that the legal regime with respect to the Area provided under LOS Convention Part XI only deals specifically with the Area's mineral resources and establishes an institutional regime based on the concept of the common heritage of mankind.⑫ The regime applies neither to non-mineral resources nor to waters superjacent to the Area.⑬ Additionally, Proelss noted that:

> **the LOS Convention, is based on the assumption that the regime of the high seas covers all activities carried out beyond the areas of national jurisdiction, irrespective of whether they are conducted in the water column or on the seabed, as long as the Convention [LOS Convention] itself does not contain any special rule to the contrary.**⑭

It has since been concluded that "there was no legal gap with respect to living resources in areas beyond national jurisdiction and that the freedoms of

---

⑩ See Tullio Treves, supra. note 497, at 16.
⑪ Ibid.
⑫ See Nele Matz-Lueck, supra note 773, at 62.
⑬ See Lyle Glowka, supra note 456, at 167.
⑭ See Alexander Proelss, supra. note 494, at 430

the high seas were applicable to activities relating to marine genetic resources"[15]. Though LOS Convention Part VII does not constitute a comprehensive and satisfactory set of legal rules, marine bioprospecting activities conducted in the Area that focus on the marine living resources or sedentary resources are subject to the regime of the high seas.

(d) Benefit Sharing Regime

Applying the freedom of the high seas principle would quite likely lead to a rush towards the exploitation of economically valuable marine genetic or biochemical resources. Some States that have the necessary technical competence to exploit the resources in ABNJ, while others States barely have the resources to provide their citizens with basic standard of living.[16] The strong will potentially get stronger, the rich richer.[17] Consequently, marine bioprospecting conducted in marine areas beyond national jurisdiction would quite likely lead to inequitable and thus unacceptable consequences. New cooperative schemes should therefore be envisaged at the international level, based on the objective benefit of all States.[18] This also corresponds with the principle of fair and equitable benefit-sharing from the utilization of genetic resources, as set out by Article 1 of the CBD.[19] The majority of commentators support an expansion of the benefit-sharing regime in Part XI of the LOS Convention to ensure that developing States can also benefit from future marine bioprospecting activities in some way.[20] In February 2006, the international regime for the genetic resources in the deep seabed was brought

---

[15] See Report of the ad hoc open-ended Informal Working Group to study issues relating to the conservation and sustainable use of marine biological diversity beyond areas of national jurisdiction, transmittal letter dated 9 March 2006 from the Co-Chair persons of the Working Group to the President of the General Assembly. UN Doc. A/61/65 (2006), para. 30.

[16] See Tullio Scovazzi, supra. note 737, at 385.

[17] Ibid.

[18] See Tullio Scovazzi, supra. note 803, at 57.

[19] Ibid.

[20] See Alexander Proelss, supra. note 494, at 438.

III. Implementing the MSR Regime in the Marine High-Tech Era

up in the Ad Hoc Open-ended Informal Working Group (Working Group). The Working Group studied issues relating to the conservation and sustainable use of marine biological diversity beyond areas of national jurisdiction and the Working Group held a second meeting in 2008.[921] During the 2008 meeting,

> some delegations proposed focusing on practical measures to enhance the conservation and sustainable use of marine genetic resources. It was proposed that such practical measures could address, among others, options for benefit-sharing. In that regard, several delegations expressed interest in considering a proposal to use the multilateral system developed under the International Treaty on Plant Genetic Resources for Food and Agriculture as a possible reference point for the discussions.[922]

Whether the benefit-sharing regime should be established as a new multilateral system or an expansion of the benefit-sharing regime under LOS Convention Part XI has thus not been consistently answered. However, it is difficult to deny that equitably sharing benefits derived from marine bioprospecting should be promoted in order to prevent developed countries from reaping a disproportionate share of the benefits and, more importantly, to help strengthen the capacity of less developed states. It is also worth nothing that the notion of "sharing benefits" itself is somewhat ambiguous. According to Nele Matz-Lueck,

> [a] sharing of benefits could range from developing countries' participation in research activities, the sharing of information on research and a transfer of technologies to process the relevant data to the sharing of economic benefits resulting from the

---

[921] See Tullio Scovazzi, supra. note 803, at 53-54.
[922] See Joint Statement of the Co-Chair Persons of the Working Group. Doc. A/63/79 of 16 May 2008, para. 32.

application of technologies developed from marine genetic resources.[123]

Whereas some provisions of the LOS Convention deliberately deal with the publication and dissemination of information, knowledge and analysis obtained from MSR,[124] and participation and capacity building of developing countries,[125] financial or other economic benefits are also stipulated in the LOS Convention.[126] To achieve a better analysis, the Authority's role should first be elaborated. The Authority is the organization through which State Parties shall organize and control activities in the Area, particularly with a view to administering the resources of the Area.[127] Since the terms "activities in the Area" and "resources of the Area" are both indicated as mineral resources-related, it can be concluded that the Authority's competence is restricted to measures in relation to the exploitation

---

[123] See Nele Matz-Lueck, supra. note 773, at 70.
[124] See LOS Convention Article 244.
[125] See LOS Convention Article 249 and Article 143 (3).
[126] See LOS Convention Article 82, 140 and 160.
[127] See LOS Convention Article 157 (1).

◆Ⅲ. Implementing the MSR Regime in the Marine High-Tech Era◆

of the Area's mineral resources.[628] Therefore, the Authority currently has no mandate to restrict or regulate marine bioprospecting activities on marine genetic or biochemical resources.[629] Similarly, LOS Convention Articles 140 (2) and 160 (2) (g) both stipulate that the Authority shall ensure the equitable sharing of financial and other economic benefits derived from activities in the Area through any appropriate mechanism. This should be done on a non-discriminatory basis, consistent with other provisions provided by the LOS Convention and Authority's rules, regulations, and procedures. De lege lata, it seems that the extent to which the benefit-sharing regime is implemented by the Authority is limited to the exploitation of non-living resources.[630]

Accordingly, the Authority's mandate does not cover benefit-sharing issue derived from marine bioprospecting activities. There have been some recent signs that the Authority is taking a closer interest in the potential environmental impacts of mining activities on marine biodiversity, while it is acknowledged that the Authority needs to know more about seabed and deep

---

[628] However, other regulatory powers granted to the Authority are still under discussion, especially with respect to marine environmental protection. Pursuant to Article 145, necessary measures shall be taken regarding activities in the Area to ensure effective protection for the marine environment from harmful effects which may arise from such activities. To this end, the Authority shall adopt appropriate rules, regulations and procedures. In this regard, on the one hand, it is observed by Tullio Scovazzi that "The regulator powers granted to the ISBA [Authority] are not limited to the harmful effects of those mining activities which belong to the typical field of competence of this organization. On the contrary, such powers are enlarged to encompass the protection and conservation of every kind of natural resource and all the fauna and flora which can be found in the Area". See Tullio Scovazzi, supra. note 737, at 393. On the other hand, it is argued that this provision only applies "with respect to activities in the Area". By referring restrictive definition of "resources" and "activities in the Area" under LOS Convention, the Authority is therefore "not legally entitled to adopt measures for the preservation of the marine environment (such as, e. g., the establishment of marine protected areas around hydrothermal vent sites), if and to the extent that these measures are not directly linked to the intention of a party (or the enterprise) to exploit the mineral resources of the respective seabed area". See Alexander Proelss, supra. note 494, at 429.

[629] See Nele Matz-Lueck, supra. note 773, at 71.

[630] See Alexander Proelss, supra. note 494, at 423.

ocean biodiversity.⁸¹ However, it also appears that the Authority wants to confine its work within its existing mandate:

> Our purpose is not to deal with it [marine biodiversity] in a comprehensive way; our purpose is to deal with it in a manner which would be of interest to the [A]uthority [in regard to the regulation of deep-sea mining]. We are not looking to control or manage or regulate marine scientific research. We are not looking to license bioprospectors or to deal with the patent rights of bio-prospectors.⁸²

Against this background, with respect to the discussion whether the mandate of the Authority should be broadened as to cover marine bioprospecting activities as well as manage a sharing of benefits or not is still under discussion. As for the former discussion, it has been noted that,

> the inclusion of deep seabed genetic resources within the regime dealing with the Area and its resources would respond to the ideas of benefit-sharing and permanent management. However, in light of the difference between mineral and biological resources, one might consider using the regime as a model rather than copying the United Nations Convention on the Law of the Sea exactly.⁸³

Regarding the latter discussion, Proelss explains that:

> [i]f one takes into account the comparably poor involvement of State Parties to UNCLOS [LOS Convention] in the work of ISA

---

⑧¹ See David Kenneth Leary, supra. note 725, at 218.

⑧² See International Seabed Authority, Press Release, 7th August, 2003, UN Doc. No. SB/9/13.

⑧³ See Marine and Coastal Biodiversity, Review, Further Elaboration and Refinement of the Programme of Work: study of the relationship between the convention on biological diversity and the United Nations Convention on the Law of the Sea with regard to the conservation and sustainable use of genetic resources on the deep seabed (decision II/10 of the Conference of the Parties to the Convention on Biological Diversity), Doc. UNEP/CBD/SBSTTA/8/INF/3/Rev. 1, para. 133.

{which comes to the fore, inter alia, in difficulties in securing the necessary participation relevant to the quorum according to Article 159 (5) UNCLOS [LOS Convention]} as well as the fact that the original approach of Part XI UNCLOS [LOS Convention] was changed radically by the provisions of the Implementation Agreement of 1995, the suggestion of expanding the deep seabed mining regime and/or the mandate of the Authority to the marine genetic resources should not be further pursued.[134]

Furthermore, in practice, the expansion of the mandate of the Authority does not seem to be a promising solution due to States' diverse political interests.[135]

As for the perspective of benefit-sharing, the issue not only involves commercial benefits, but is always intimately connected with intellectual property rights such as patents.[136] Unlike the mineral resources of the Area, where the primary issues often relate to engineering and commercial feasibility, the commercial exploitation of the deep seabed's genetic and biochemical resources is a far more complicated and lengthy process. It not only entails the simple processes of digging something up, bringing it to the surface, and processing it for sale. Rather, the reward is also the monopoly of this practice that comes with the grant of a patent, and the associated profit that flows from its exploitation.[137] Apparently, the Authority does not have any expertise in the patents, particularly patents associated with biotechnology.[138] Furthermore, the Authority has no expertise in coordinating and distributing funds either, such as patent royalties. In spite of that under LOS Convention Article 160 (1) (f) the Authority is required to develop rules, regulations, and procedures for the equitable sharing of financial and

---

[134] See Alexander Proelss, supra. note 494, at 440.
[135] See Nele Matz-Lueck, supra. note 773, at 72.
[136] See David Kenneth Leary, supra. note 725, at 222.
[137] Ibid.
[138] Ibid.

other economic benefits derived from mining activities in the Area, and pursuant to LOS Convention Article 82, the Authority is responsible for distributing the proceeds derived from coastal States exploiting non-living resources of the continental shelf beyond 200 nautical miles. That said, to date the Authority has not developed any rules, regulations, or procedures for the equitable sharing of financial or other economic benefits derived from mining activities in the Area, nor has it disbursed any proceeds from the exploitation of non-living resources on the outer continental shelf.[39] The Authority's ability to manage the disbursement process is therefore unproven.[40] It is more appropriate that, pursuant to LOS Convention Article 143, the Authority should focus on disseminating knowledge about the genetic or biochemical resources in the deep seabed and, promoting and facilitating MSR.[41] In large part, these activities are already undertaken within the terms of the Authority's existing mandate.[42] More fundamentally, while other than financial or other economic benefits, these activities are resulting in some forms of benefit-sharing as well. It is therefore questionable whether it is appropriate to expand the Authority's mandate to encompass the benefit-sharing regime, especially as far as financial or other economic benefits are concerned, for the exploitation of marine genetic or biochemical resources in the Area other than mineral resource mining. More importantly, given its existing structure and expertise, it seems to be an organization not too well-suited to play such a broader role.[43]

Having established that expanding the Authority's mandate in terms of financial benefit-sharing is not a viable solution, creating a new instrument does not seem to be a promising approach either. According to Nele Matz-

---

[39] See David Kenneth Leary, supra. note 725, at 179.
[40] Ibid.
[41] Ibid., at 223.
[42] Ibid.
[43] Ibid., at 224.

Lueck, not only does "the proliferation of treaties and other binding or non-binding regulations on different levels bears its own difficulties",⁴⁴ but "States have[also] become particularly reluctant to transfer powers and financial resources to a growing number of international organizations or institutions necessary to operate modern treaty regimes".⁴⁵ It follows that the attention should be devoted to "effectively implementing and carefully amending the existing rules"⁴⁶ to achieve an equitable benefit-sharing regime. In this vein, the LOS Convention refers to cooperation through "competent international organizations" on many occasions, especially living resource conservation, marine environmental protection, and the conduct of MSR. Competent international organizations that have the expertise and experience to deal with disbursing large sums of money and more importantly, that have substantial experience in funding or managing projects associated with sustainable development, would be the most desirable choice.⁴⁷ However, as mining activities in the Area may harm ecosystems with particular genetic or biochemical resources, experience in integrated mining management and genetic and biochemical resource conservation is also indispensable.⁴⁸ Regardless, any international organization that is responsible for the management of benefit-sharing schemes, as well as other aspects originated from marine bioprospecting activities, would have to closely cooperate with the Authority in order to avoid confusion and to coordinate activities.⁴⁹

Regarding potential benefit-sharing schemes, ocean development taxes, as advocated by Elizabeth Mann Borgese, and a global commons trust fund are the two main alternatives, although both of them are still subject to

---

⑭ See Nele Matz-Lueck, supra. note 773, at 64.
⑮ Ibid., at 71.
⑯ See Alexander Proelss, supra. note 494, at 444.
⑰ See David Kenneth Leary, supra. note 725, at 179-180.
⑱ See Nele Matz-Lueck, supra. note 773, at 71.
⑲ Ibid., at 72.

199

various divergences and have been canvassed at length in the literatures.[650] Another important element that would highly affect any benefit-sharing scheme is intellectual property rights resulting from patents over health-related inventions from marine bioprospecting activities. Without delving any further into an analysis of the collection or disbursement of those funds or fees, it is noteworthy that intellectual property rights, principally patents, are clearly a key economic incentive behind marine bioprospecting.[651] Appropriately dealing with these rights might be effectively channeled for the benefit of developing countries.[652] In this respect, David Kenneth Leary suggested that "the resources could be shared by means of royalties payable in relation to patents granted in relation to deep-sea genetic resources"[653]. However, this suggestion would only apply to the genetic or biochemical resources located beyond national jurisdiction, since any resources within national jurisdiction are now firmly entrenched by the international law.[654]

Moreover, other than financial or economic benefits, further considerations are necessary. According to the 2010 recommendations of the Working Group, a number of practical measures that address implementation gaps in relation to marine genetic resources in ABNJ were highlighted, such as promotion of MSR; the development of codes of conduct for research activities; environmental impact assessments; the establishment of mechanisms for cooperation and the sharing of information and knowledge resulting from research on marine genetic resources; and so on.[655]

---

[650] More information and discusses about ocean development taxes and global commons trust see David Kenneth Leary, supra. note 725, at 176-178.
[651] Ibid., at 175.
[652] See Alexander Proelss, supra. note 494, at 446.
[653] See David Kenneth Leary, supra. note 725, at 175.
[654] Ibid.
[655] Letter dated 16 March 2010 from the Co-Chairs of the ad hoc open-ended Informal Working Group to the President of the General Assembly, UN Doc. A/65/68, 17 March 2010, para. 73.

◇ Ⅲ. Implementing the MSR Regime in the Marine High-Tech Era ◇

Having said that, a further gap exists in respect of patent application that ought to be addressed. Only a limited number of technologically advanced States have the ability to conduct marine bioprospecting in the international marine area, and some of them have already taken out related patents that ensure patentees all the benefits connected with the commercialization of the patented substances. This runs counter to the underlying principles of the LOS Convention's aim to establish a just and equitable international legal order and promote the utilization and conservation of marine resources on a sustainable basis.[856] According to a 2004 report by the United Nations Secretary-General,

> [a]s legitimate as the protection of private data and proprietary interests through intellectual property rights may be, a balance needs to be struck between private benefits and benefits to humankind as a whole through the advancement of scientific knowledge.[857]

Furthermore, since the degree of confidentiality required prior to filing for patents is neither compatible with the requirement for disseminating and publishing data and research results,[858] nor the equitable benefit-sharing rationale, the patent framework could therefore be complemented by a tool called open-source licensing.[859] Contrary to experimental use exemptions,[860] open-source licensing generally does not entail non-commercial characteristic as prerequisite.[861] The applicability of open-source licensing would thus not

---

[856] See Tullio Scovazzi, supra. note 737, at 406.

[857] See Report of the Secretary-General on Oceans and the Law of the Sea, 4 March 2004. UN Doc A/59/62, para. 260.

[858] See Report of the Secretary-General, Oceans and the Law of the Sea, UN Doc. A/62/66/Add. 2 (2007), para. 241.

[859] See Alexander Proelss, supra. note 494, at 446.

[860] Experimental use exemptions allow scientists to use a patented invention, provided that the research is for non-commercial purposes. See Report of the Secretary-General, Oceans and the Law of the Sea, UN Doc. A/62/66 (2007), para. 227.

[861] See UN Doc. A/62/66/Add. 2 (2007), supra. note 858, footnote 109.

just reflect the emphasis on effectively disseminating the results and analysis of MSR conducted in the Area, as stipulated by LOS Convention Article 143 (3) (c). Rather, it would also further strengthen the implementation of the equitable benefit-sharing idea in non-financial or economic respects.

4. Conclusion

Four decades have now passed since hydrothermal vents, with their abundant biodiversity and biotechnological potential, were discovered. However, the international community has not yet adopted any definitive proposal for a regime to govern marine bioprospecting.[62] The absence of MSR and bioprospecting definitions from either the LOS Convention or the CBD has inevitably frustrated any endeavour to develop a regime for regulating marine bioprospecting and ensure the sustainable exploitation of marine genetic and biochemical resources. Given that it is difficult to distinguish marine natural resources exploration from MSR, marine bioprospecting may be considered as applied MSR with the nature of exploring marine natural resources. As far as jurisdictional issues are concerned, within coastal State's internal waters and territorial sea, due to the nature of exploration, marine bioprospecting is fully subject to the coastal State's sovereignty. Similarly, pursuant to LOS Convention Article 245, coastal State's sovereignty applies as well. Within coastal State's EEZ or on the continental shelf, due to that sovereign rights enjoyed by the coastal State are superior to its jurisdiction, marine bioprospecting conduct in the EEZ or on the continental shelf is subject to the sovereign rights of coastal State, rather than the jurisdiction as provided in LOS Convention Article 246. The regulation of marine bioprospecting activities in ABNJ, which has been on the agenda of various international bodies for over a decade with little concrete progress,[63] is far more complicated because of the Authority's in-

---

[62] See Robin Warner, supra. note 679, at 413.
[63] Ibid.

volvement and the "common heritage of mankind" concept. Instead of spending the next couple of decades on divisive discussions that will not likely yield a convincing regime, the international community should focus more on increasing opportunities for capacity-building through either benefit-sharing regimes or other initiatives.[864] With respect to the notifying regime proposed in the case of VOS,[865] it does not seem appropriate to apply it to marine bioprospecting. Marine bioprospecting is not pure MSR, rather bears great relevance with the exploration of marine natural resources and the potential prospect of economic benefit from marine bioprospecting that could be huge. In the view of protecting coastal State's interests and avoiding emergence of inequitable, unacceptable consequences, notifying regime should not apply in marine bioprospecting.

Equitable benefit-sharing, a key approach to supporting sustainable development, is not just a form of financial or economic benefit-sharing, but more importantly a tool for capacity-building and sharing technology. It is suggested that practical mechanisms and options for benefit-sharing should include addressing monetary and non-monetary benefits for equitable distribution.[866] It appears that most developing countries lack the infrastructure and expertise to perform marine bioprospecting.[867] In order for developing countries to develop the necessary capacities to fully benefit from marine genetic or biochemical resources, sharing research and development results, participating in scientific research, development programmes, accessing databases, having ex situ biological inventories on preferential terms, and promoting institutional capacity-building are some examples of what is

---

[864] See Margaret F. Hayes, supra. note 500, at 700.
[865] See supra. Chapter III, B (2) (c).
[866] See Letter dated 8 June 2012 from the Co-Chairs of the ad hoc open-ended Informal Working Group to the President of the General Assembly, UN Doc. A/67/95, supra. note 746, para. 19.
[867] See UN Doc. A/62/66/Add. 2 (2007), supra. note 858, para. 244.

required.[68] The experience gained from International Treaty on Plant Genetic Resources for Food and Agriculture and the future implementation of the Nagoya Protocol,[69] could be considered.[70]

Additionally, it should not be neglected that the overall goal of the international community should be the conservation and sustainable use of marine biodiversity. The "first come, first served" approach existing on the high seas and arguably, in the Area is counterproductive and undermines sustainability.[71] In that regard, it is noted, "the common heritage of mankind was not solely about benefit sharing, but just as much about conservation and preservation".[72] Against this background, the need to consider the role of area-based management tools, including marine protected areas, to support ecosystem approaches at the national, regional and global levels, is highlighted.[73]

---

[68] See UN Doc. A/62/66/Add. 2 (2007), supra. note 858, para. 246.

[69] The Nagoya Protocol on Access to Genetic Resources and the Fair and Equitable Sharing of Benefits Arising from their Utilization is a supplementary agreement to the CBD. Its objective is the fair and equitable sharing of benefits arising from the utilization of genetic resources, including by appropriate access to genetic resources and by appropriate transfer of relevant technologies, taking into account all rights over those resources and to thechnologie, and by appropriate funding, thereby contributing to the conservation of biological diversity and the sustainable use of its components. More information is available at http://www.cbd.int/abs/about/ (last visited on 13 August 2013).

[70] See Letter dated 8 June 2012 from the Co-Chairs of the ad hoc open-ended Informal Working Group to the President of the General Assembly, UN Doc. A/67/95, supra. note 746, para. 18.

[71] See Letter dated 30 June 2011 from the Co-Chairs of the ad hoc open-ended Informal Working Group to the President of the General Assembly, UN Doc. A/66/119, 30 June 2011, para. 17.

[72] Ibid.

[73] See Letter dated 8 June 2012 from the Co-Chairs of the ad hoc open-ended Informal Working Group to the President of the General Assembly, UN Doc. A/67/95, supra. note 746, para. 20.

Ⅳ. Final Remarks

The LOS Convention regulates MSR, but it does not define it. Although the impact of the MSR regime stipulated by the LOS Convention on the marine science development is unclear, the regime seems to be working without significant problems.[674] Annually, hundreds of requests for consent to conduct MSR projects in areas under coastal States' jurisdiction are made, and very few of these requests are rejected or delayed.[675]

In terms of MSR under the LOS Convention, many efforts have been made to strike a balance between the interests of coastal States and the marine scientific community. While the LOS Convention actively promotes the flow of scientific data and information and the transfer of knowledge resulting from MSR,[676] it still should be noted that the prior approval for making the results internationally available should also be required if the results are of direct significance to exploring and exploiting natural resources.[677]

It follows from the foregoing that we know much more about the ocean environment today, that we have discovered a range of major new problems, and that we have much better tools for conducting MSR.[678] Not all of these outcomes were foreseen when the LOS Convention was adopted. There continues to be unfinished business from UNCLOS Ⅲ, since it is expected that a legal regime built on science

---

[674] See A. H. A. Soons, supra. note 664, at 162.
[675] Ibid.
[676] See LOS Convention, Article 244.
[677] See LOS Convention, Article 249.
[678] See Aldo Chircop, supra. note 27, at 606.

and technology will experience changes on an ongoing basis.[79] Now the challenge is to see to what extent the underlying scientific and technological assumptions in the LOS Convention are still valid in light of today's more sophisticated and capable scientific research.[80] It is essential to continue these discussions not only among scholars and policy makers, but also with the wider marine scientific community.[81] The interests of the scientific community are mainly represented by a number of international scientific organizations, namely the Intergovernmental Oceanographic Commission (IOC),[82] the International Hydrographic Organization (IHO),[83] the International Council for Science (ICSU),[84] and the North Pacific Marine Science Organization (PICES).[85] Since the adoption of the LOS Convention, these organizations, especially the IOC, have dealt with the implementation of the MSR regime according to various developments. Most importantly, in 1997, the IOC Assembly established a new standing subsidiary body, the Advisory Body of Experts on the Law of the Sea (IOC/ABE-LOS).[86] The Advisory Body is tasked with providing advice on the IOC's role in relation to the LOS Convention.[87] When the issue arose of how the MSR regime applies to floats and buoys deployed on the high seas but which drifted into the EEZ, the IOC Assembly was requested to ask the ABE-LOS to provide practical guidelines.[88] It is therefore submitted that the involvement of competent international marine scientific organizations in im-

---

[79] See Aldo Chircop, supra. note 27, at 614.
[80] Ibid., at 582.
[81] See A. H. A. Soons, supra. note 664, at 163.
[82] See the website http://ioc-unesco.org/ (last visited on 12 April 2013).
[83] See the website http://www.iho.int/srv1/ (last visited on 12 April 2013).
[84] See the website http://www.icsu.org/ (last visited on 12 April 2013).
[85] See the website http://www.pices.int/ (last visited on 12 April 2013).
[86] See A. H. A. Soons, supra. note 664, at 150.
[87] Several topics the IOC/ABE-LOS has been working on see http://ioc-unesco.org/index.php?option=com_content&view=article&id=364&Itemid=100047 (last visited on 12 April 2013).
[88] See Aldo Chircop, supra. note 27, at 614. Also see Katharina Bork, etc., supra. note 450, at 317.

◆ Ⅳ. Final Remarks ◆

plementing the MSR regime is critical since the role these organizations play in building knowledge networks developing guidelines to address existing and future gaps is indispensable.

Bibliography

Aldous, Don: New Technology in Surveillance and Enforcement and Appropriate Development in the Pacific Islands, in Lewis M. Alexander/Scott Allen/Lynne Carter Hanson (eds.), New Development in Marine Science and Technology: Economic, Legal and Political Aspects of Change, The Law of the Sea Institute, University of Hawaii, 1988, S 153-170.

Allot, Philip: The Concept of International Law, EJIL, 10 (1999), S 31-50.

—. Making the New International Law: Law of the Sea as Law of the Future, IJ, 40 (1984/1985), S 442-460.

Anderson, A. W. : Jurisdiction over Stateless Vessels on the High Sea: An Appraisal under Domestic and International Law, in Journal of Maritime Law and Commerce, 13 (1982), S323-336.

Anderson, David: Scientific Evidence in Cases under Part XV of the LOSC, in Myron H. Nordquist/Ronan Long/Tomas H. Heidar/John Norton Moore (eds.), Law, Science & Ocean Management, Martinus Nijhoff Publishers, 2008, S505-518.

—. Freedoms of the High Seas in the Modern Law of the Sea, in David Freestone/ Richard Barnes/ David M. Ong (eds.), The Law of the Sea, Progress and Prospects, S 327-346.

Anderson, Lee: Is the LOS Convention Addressing Today's Changing Marine Scientific, Technological, Economic, Legal and Political Issues? The Fishing Perspective, in Lewis M. Alexander/Scott Allen/Lynne Carter Hanson (eds.), New Development in Marine Science and Technology: Economic, Legal and Political Aspects of Change, The Law of the Sea Institute, University of Hawaii, 1988, S 460-465.

Bateman, Sam: Hydrographic Surveying in Exclusive Economic Zones—Is it Marine Scientific Research? in Myron H. Nordquist/Tommy Koh/John Norton Moore (eds.), Freedom of Seas, Passage Rights and The 1982 Law of the Sea Convention, Martinus Nijhoff Publishers, 2009, S 105-131.

Becher, Michael A. : The Shifting Public Order of the Oceans: Freedom of Navigation and the Interdiction of Ships at Sea, HILJ, 46 (2005), S 131-230.

Belsky, Martin H.: Marine Ecosystem Model: The Law of the Sea's Mandate for Comprehensive Management, in Lewis M. Alexander/Scott Allen/Lynne Carter Hanson (eds.), New Development in Marine Science and Technology: Economic, Legal and Political Aspects of Change, The Law of the Sea Institute, University of Hawaii, 1988, S 115-134.

Birnie, Patricia W.: Law of the Sea and Ocean Resources: Implications for Marine Scientific Research, IJMCL, 10 (1995), S 229-252.

—. New Technologies: Effects on the Developing Law of Fisheries, in Jean-Pierre Beurier/Alexandre Kiss/Said Mahmoudi (eds.), New Technologies and Law of the Marine Environemnt, Kluwer Law International, 2000, S 23-39.

Birnie, Patricia W. /Boyle, Alan E. /Redgwell, Catherine: International Law and the Environment, Oxford, 2009.

Boczek, Boleslaw A.: Peaceful Purposes Provisions of the United Nations Convention on the Law of the Sea, ODIL, 20 (1989), S 359-389.

Bodansky, Daniel: The Legitimacy of International Governance: A Coming Challenge for International Environment Law?, AJIL, 93 (1999), S 596-624.

—. May We Engineer the Climate, Climate Change, 33 (1996), S 309-321.

Bork, Katharina et al.: The Legal Regulation of Floats and Gliders—In Quest of a New Regime? Ocean Development and International Law, 39 (2008), S 298-328.

—. Der Rechtsstatus von unbemannten ozeanographischen Messplattformen im international Seerecht, Baden-Baden: Nomos 2011.

Bothe, Michael: Measures to Fight Climate Change—A Role for the Law of the Sea, in Holger Hestermeyer/Nele Matz-Lueck/Anja Seibert-Fohr/Silja Voeneky (eds.), Law of the Sea in Dialogue, Springer 2010, S 31-46.

Boyle, Alan: The Environment Jurisprudence of the International Tribunal for the Law of the Sea, IJMCL, 22 (2007), S 369-381.

Brown, E. D.: Freedom of Scientific Research and the Legal Regime of Hydrospace, IJIL, 9 (1969), S 327-380.

—. The International Law of the Sea, Vol. I, Dartmouth Publishing, 1994.

—. The Significance of a Possible EC EEZ for the Law Relating to Artificial Islands, Installations, and Structures, and to Cables and Pipelines, in the Exclusive Economic Zone, ODIL, 23 (1992), S 115-144.

Burger, W.: Treaty Provisions Concerning Marine Science Research, ODIL, 1 (1973), S 159-184.

Burke, William T. : Highly Migratory Species in the New Law of the Sea, ODIL, 14 (1984), S273-314.

Caflisch, Lucius/Piccard, Jacques: The Legal Regime of Marine Scientific Research and the Third United Nations Conference on the Law of the Sea, Zeitschrift für ausländisches öffentliches Recht und Völkerrecht, 38 (1978), S 848-901.

Caldeira, Ken et al. : Depth, Radiocarbon, and the Effectiveness of Direct CO_2 Injection as an Ocean Carbon Sequestration Strategy, Geophysical Research Letters, 29 9 (2002), S 1-4.

Capt. Roach, J. Ashley: Defining Scientific Research: Marine Data Collection, in Myron H. Nordquist/Ronan Long/Tomas H. Herdar/John Norton Moore (eds.), Law, Science & Ocean Management, Martinus Nijhoff Publishers 2008, S 541-574.

—. Marine Data Collection: Methods and the Law, in Myron H. Nordquist/Tommy Koh/John Norton Moore (eds.), Freedom of Seas, Passage Rights and The 1982 Law of the Sea Convention, Martinus Nijhoff Publishers, 2009, S 171-208.

Carlson, Eric: Coastal and Ocean Planning, in Lewis M. Alexander/Scott Allen/Lynne Carter Hanson (eds.), New Development in Marine Science and Technology: Economic, Legal and Political Aspects of Change, The Law of the Sea Institute, University of Hawaii, 1988, S 388-389.

Charlier, Roger H./ Charlier, Constance C. : Ocean Non-Living Resources: Historical Respective on Exploitation, Economics and Environmental Impact, in International Journal of Environmental Studies, 40 (1992), S 123-134.

Chircop, Aldo: Advances in Ocean Knowledge and Skill: Implications for the MSR Regime, in Myron H. Nordquist/Ronan Long/Tomas H. Heidar/John Norton Moore (eds.), Law, Science & Ocean Management, Martinus Nijhoff Publishers, 2008, S575-615.

Churchill, R. R./ Lowe, A. V. : The Law of the Sea, Manchester University Press, 1999.

Clingan, Thomas A. : The Law of the Sea: Ocean Law and Policy, Austin&Winfield San Francisco-London, 1994.

Corell, Robert: Marine Science in the 1990s: Global Change and Its Implications, in Lewis M. Alexander/Scott Allen/Lynne Carter Hanson (eds.), New Development in Marine Science and Technology: Economic, Legal and Political Aspects of Change, The Law of the Sea Institute, University of Hawaii, 1988, 27-33.

Corner, Adam/Pidgeon, Nick F. : Geoengineering the Climate: The Social and Ethical Implications, Environment: Science and Policy for Sustainable Development, 52 (2010), S

24-37.

Craven, John P.: Technology and the Law of the Sea (A Twenty Year Reprise) in Lewis M. Alexander/Scott Allen/Lynne Carter Hanson (eds.), New Development in Marine Science and Technology: Economic, Legal and Political Aspects of Change, The Law of the Sea Institute, University of Hawaii, 1988, S 36-49.

Dai, Ruguang: Management of Oceanographic Data and International Cooperation in China, in Lewis M. Alexander/Scott Allen/Lynne Carter Hanson (eds.), New Development in Marine Science and Technology: Economic, Legal and Political Aspects of Change, The Law of the Sea Institute, University of Hawaii, 1988, S 342-344.

Danilenko, Gennady M.: The Theory of International Customary Law, GYIL, 31 (1988), S 9-47.

Denman, Kenneth L.: Climate Change, Ocean Processes and Ocean Iron Fertilization, MEPS, 364 (2008), S 219-225.

Dickinson, Robert E.: Climate Engineering: A Review of Aerosol Approaches to Changing the Global Energy Balance, Climatic Change, 33 (1996), S 279-290.

Dzidzornu, David M.: Four Principles in Marine Environment Protection, ODIL, 29 (1998), S 91-123.

Ehlers, Peter: The Intergovernmental Oceanographic Commission: An International Organisation for the Promotion of Marine Research, IJMCL, 15 (2000), S 533-554.

Elferink, Alex G. Oude: The Regime of the Area: Delineating the Scope of Application of the Common Heritage Principle and Freedom of the High Seas, IJMCL, 22 (2007) S 143-176.

Franssen, Herman T.: Developing Country Views of Sea Law and Marine Science, in Warren S. Wooster (ed.), Freedom of Oceanic Research: A Study Conducted by the Center for Marine Affairs of the Scripps Institution of Oceanography, Crane, Russak&Company, Inc. 1973, S 137-177.

Garrison, Tom: Oceanography: An Invitation to Marine Science, Belmont, 2010.

Gavouneli, Maria: Functional Jurisdiction in the Law of the Sea, Martinus Nijhoff Publishers, 2007.

Glowka, Lyle: The Deepest of Ironies: Genetic Resources, Marine Scientific Research, and the Area, Ocean Yearbook, 12 (1996), S 154-178.

Godin, Raymond H.: Developing Technologies within the Oceanographic Component of the World Climate Research Program, in Lewis M. Alexander/ScottAllen/Lynne Carter Hanson (eds.), New Development in Marine Science and Technology: Economic, Legal and Political Aspects of Change, The Law of the Sea Institute, University of Hawaii,

1988, S 337-341.

Gold, Edgar: International and the Law of the Sea: New Directions: For a Traditional Use?, in Lewis M. Alexander/Scott Allen/Lynne Carter Hanson (eds.), New Development in Marine Science and Technology: Economic, Legal and Political Aspects of Change, The Law of the Sea Institute, University of Hawaii, 1988, S 200-213.

Gorina-Ysern, Montserrat: An International Regime for Marine Scientific Research, Transnational Publishers, Inc. 2003.

Greiveldinger, Geoffrey: Is the LOS Convention Addressing Today's Changing Marine Scientific, Technological, Economic, Legal and Political Issues? The Militray Perspective, in Lewis M. Alexander/Scott Allen/Lynne Carter Hanson (eds.), New Development in Marine Science and Technology: Economic, Legal and Political Aspects of Change, The Law of the Sea Institute, University of Hawaii, 1988, S 442-447.

Grunawalt, Richard J.: New Legal Issues Resulting from the US Global Military Commitment: A Naval Perspective of the Persian Gulf Tank War, in Lewis M. Alexander/Scott Allen/Lynne Carter Hanson (eds.), New Development in Marine Science and Technology: Economic, Legal and Political Aspects of Change, The Law of the Sea Institute, University of Hawaii, 1988, S 229-233.

Gurun, Gwénaëlle Le: EIA and the International Seabed Authority, in Kees Bastmeijer and Timo Koivurova (eds.), Theory and Practice of Transboundary Environmental Impact Assessment, Brill, 2008, S 221-263.

Hayashi, Moritaka: Military and Intelligence Gathering Activities in the EEZ: Definition of Key Terms, Marine Policy, 29 (2005), S 123-137.

Hayes, Margaret F. : Charismatic Microfauna: Marine Genetic Resources and the Law of the Sea, in Myron H. Nordquist, Ronan Long, Tomas H. Heidar and John N. Moore (eds.), Law, Science & Ocean Management, Martinus Nijhoff, Publishers, 2008, S 683-700.

Ijlstra, T.: Implementation of the United Nations Law of the Sea Convention: Removal of Offshore Installations, in Budislav Vukas (ed.), Essays on the New Law of the Sea 2, Zagreb, 1990, S 55-79.

Jones, Erin Bain: Law of the Sea Oceanic Resources, Southern Methodist University Press 1972.

Kildow, Judith A. Tegger: Nature of the Present Restrictions on Oceanic Research, in Warren S. Wooster (ed.), Freedom of Oceanic Research: A Study Conducted by the Center for Marine Affairs of the Scripps Institution of Oceanography, Crane, Russak&Company, Inc. 1973, S 5-28.

Kirchner, Andree: Bioprospecting, Marine Scientific Research and the Patentability of Genetic Resources, in Norman A. Martínez Gutiérrez (ed.), Serving the Rule of International Maritime Law, Routledge, 2010, S 119-128.

Kirke, Brian: Enhancing Fish Stocks with Wave-Powered Artificial Upwelling, Ocean and Coastal Management, 46 (2003), S 901-905.

Kiss, Alexandre/Shelton, Dinah: International Environmental Law, Ardsley, 2000.

Knauss, John A.: Development of the Freedom of Scientific Research Issue of the Third Law of the Sea Conference, ODIL, 1 (1973), S 93-120.

—. Recent Experience of the United States in Conducting Marine Scientific Research in Coastal State Exclusive Economic Zones, in Thomas A. Clingan (ed.), The Law of the Sea: What Lies Ahead, the Law of the Sea Institute, University of Hawaii, S 297-309.

Laursen, Finn: Toward a New International Marine Order, Martinus Nijhoff Publishers, 1982.

Leanza, U.: Scientific Research and the Law of the Sea, in Budislav Vukas (ed.), Essays on the New Law of the Sea 2, ZAGREB, 1990, S 81-107.

Leary, David Kenneth: International Law and the Genetic Resources of the Deep Sea, Martinus Nijhoff Publishers, 2007.

Lefeber, Rene: Creative Legal Engineering, LJIL, 13 (2000), S 1-9.

Liang, Nai-Kuang/ Peng, Hai-Kuen: A Study of Air-Lift Artificial Upwelling, Ocean Engineering, 32 (2005), S 731-745.

Lin, Albert C.: Geoengineering Governace, Issues in Legal Scholarship, 8 (2009), S 1-24.

Long, Ronán: Regulating Marine Biodiscovery in Sea Areas under Coastal State Jurisdiction, in Myron H. Nordquist/Tommy Koh/John Norton Moore (eds.), Freedom of Seas, Passage Rights and The 1982 Law of the Sea Convention, Martinus Nijhoff Publishers, 2009, S 133-169.

Lovelock, James E./Rapley, Chris G.: Ocean Pipes Could Help the Earth to Cure Itself, available at: www.nature.com/nature/journal/v449/n7161/full/449403a.html (last visited 15th April 2013).

Lueger, Heike/Körtzinger, Arne et al.: CO_2 Fluxes in the Subtropical and Subarctic North Atlantic Based on Measurements from a Volunteer Observing Ship, Journal of Geophysical Research, 111 (2006).

MacDonald, John M.: Artifical Reef Debate: Habitat Enhancement or Waste Disposal, ODIL, 25 (1994), S 87-118.

Maes, Frank: The International Legal Framework for Marine Spatial Planning,

Marine Policy, 32 (2008), S 797-810.

Malone, Thomas: A New Dimension of International and Interdisciplinary Cooperation, in Lewis M. Alexander/Scott Allen/Lynne Carter Hanson (eds.), New Development in Marine Science and Technology: Economic, Legal and Political Aspects of Change, The Law of the Sea Institute, University of Hawaii, 1988, S 17-24.

Maruyama, Shigenao: Artificial Upwelling of Deep Seawater Using the Perpetual Salt Fountain for Cultivation of Ocean Desert, Journal of Oceanography, 60 (2004), S 563-568.

Matz-Lueck, Nele: Marine Biological Resources: Some Reflections on Concepts for the Protection and Sustainable Use of Biological Resources in the Deep Sea, Non-State Actors and International Law, 2 (2002), S 279-300.

—. The Concept of the Common Heritage of Mankind: Its Viability as a Management Tool for Deep-Sea Genetic Resources, in Erik J. Molenaar, Alex G. Oude Elferink (eds.), The International Legal Regime of Areas Beyond National Jurisdiction: Current and Future Developments, Martinus Nijhoff, 2010, S 61-75.

Mayer, Larry: Arctic Marine Research: The Perspective of a US Practitioner, in Susanne Wasum-Rainer/Ingo Winkelmann/Katrin Tiroch (eds.), Arctic Science, International Law and Climate Change, Springer, 2011, S 83-96.

McDorman, Ted L.: The Entry into Force of the 1982 LOS Convention and the Article 76 Outer Continental Shelf Regime, IJMCL, 10 (1995), S 165-187.

McDorman, Ted L./Bolla, Alexander J./Johnston, Douglas M./Duff, John: International Ocean Law: Materials and Commentaries, Carolina Academic Press, 2005.

Mensah, Thomas: Is the LOS Convention Addressing Today's Changing Marine Scientific, Technological, Economic, Legal and Political Issues? The Shipping Perspective, in Lewis M. Alexander/Scott Allen/Lynne Carter Hanson (eds.), New Development in Marine Science and Technology: Economic, Legal and Political Aspects of Change, The Law of the Sea Institute, University of Hawaii, 1988, S 448-459.

Menzel, Eberhard: Scientific Research on the Sea-Bed and Its Regime, in Symposium on the International Regime of the Sea-Bed, Instituto Affari Internazionali, 1969, S 619-647.

Mgbeoji, Ikechi: (Under) Mining the Seabed? Between the International Seabed Authority Mining Code and Sustainable Bioprospecting of Hydrothermal Vent Ecosystems in the Seabed Area: Taking Precaution Seriously, Ocean Yearbook, 18 (2004), S 413-452.

Mukherjee, P. K.: The Consent Regime of Oceanic Research in the New Law of the Sea, Marine Policy, 5 (1981), S 98-113.

Nixdorf, Uwe: Arctic Research in Practice, in Susanne Wasum-Rainer/Ingo Winkelmann/Katrin Tiroch (eds.), Arctic Science, International Law and Climate Change, Springer, 2011, S 67-81.

Nordquist, Myron H.: United Nations Convention on the Law of the Sea 1982: A Commentary, Vol. I, Martinus Nijhoff Publishers, 1989.

Nordquist, Myron H./Nandan, Satya N./Rosenne, Shabtai (eds.): United Nations Convention on the Law of the Sea 1982: A Commentary, Vol. II, Martinus Nijhoff Publishers, 1993.

—. United Nations Convention on the Law of the Sea 1982: A Commentary, Vol. III, Martinus Nijhoff Publishers, 1989.

Nordquist, Myron H./Rosenne, Shabtai/Yankov, Alexander (ed.): United Nations Convention on the Law of the Sea 1982: A Commentary, Vol. IV, Martinus Nijhoff Publishers, 1991.

Ogiwara, Seiko et al.: Conceptual Design of a Deep Ocean Water Upwelling Structure for Development of Fisheries, in Jin S. Chung and Valcana Stoyanove (eds.), Proceedings of the Fourth (2001) Ocean Mining Symposium, Cupertino: International Society of Offshore and Polar Engineers, 2001, S150-157.

Oschlies, Andreas et al.: Climate Engineering by Artificial Ocean Upwelling: Channelling the Sorcerer's Apprentice, Geophysical Research Letters, 37 (2010) S 1-5.

Ouchi, Kazuyuki: Outline of the Ocean Nutrient Enhancer "TAKUMI", Journal of Ocean Science and Technology, 2 (2005) S 9-15.

Oxman, Bernard H.: The Third United Nations Conference on the Law of the Sea: The 1977 New York Session, AJIL, 72 (1978), S 57-83.

Papadakis, Nikos: Artificial Islands, Installations and Structures in the Exclusive Economic Zone, in B. Conforti et al. (eds.), La Zona Economica Esclusiva, Milano, 1983, S 99-114.

—. The International Legal Regime of Artificial Islands, Leyden, 1977.

Proelss, Alexander: The Law on the Exclusive Economic Zone in Perspective: Legal Status and Resolution of User Conflicts Revisited, in Aldo Chircop, et al. (eds.), Ocean Yearbook, 26 (2012), S 87-112.

—. The Role of the Authority in Ocean Governance, David D. Caron and Harry Scheiber (eds.), Institutions and Regions in Ocean Governance, Martinus Nijhoff Publishers, 2012.

—. Marine Genetic Resources under UNCLOS and the CBD, German Yearbook of International Law, 51 (2008), S 417-446.

Proelss, Alexander/Guessow, Kerstin: Carbon Capture and Storage from the Perspective of International Law, EYIEL, 2 (2011), S 151-168.

Proelss, Alexander/Krivickaite, Monika: Marine Biodiversity and Climate Change, CCLR, (2009), S 437-445.

Purdy, Ray/Macrory, Richard: Geological Carbon Sequestration: Critical Legal Issues, Tyndall Center Working Papers No. 54, January, 2003.

Ramakrishna, Kilaparti/Bowen, Robert E./Archer, Jack H.: Outer Limits of Continental Shelf—A Legal Analysis of Chilean and Ecuadorian Island Claims and US response, Marine Policy, 11 (1987), S 58-68.

Rasool, S. I.: Maximizing the Benefits from the New Technologies of Oceanographic Data Gethering and Management, in Lewis M. Alexander/Scott Allen/Lynne Carter Hanson (eds.), New Development in Marine Science and Technology: Economic, Legal and Political Aspects of Change, The Law of the Sea Institute, University of Hawaii, 1988, S 333-336.

Rauch, Elmar: Military Uses of the Ocean, German Yearbook of International Law, 28 (1985), S 229-258.

Ré, Pedro: Deep-Sea Hydrothermal Vents "Oases of the Abyss", in Jean-Pierre Beurier, Alexander Kiss and Said Mahmoudi (eds.) New Technologies and Law of the Marine Environment, Kluwer Law International, 2000, S 67-74.

Redfield, Michael: The Legal Framework for Oceanic Research, in Warren S. Wooster, (ed.), Freedom of Oceanic Research: A Study Conducted by the Center for Marine Affairs of the Scripps Institution of Oceanography, Crane, Russak&Company, Inc., 1973, S 41-96.

Revelle, Roger: Scientific Research on the Sea-BedĒInternational Cooperation in Scientific Research and Exploration of the Sea-Bed, in Symposium on the International Regime of the Sea-Bed, Instituto Affari Internazionali, 1969, S 649-663.

Ringeard, Gisele: Scientific Research: from Freedom to Deontology, ODIL,1(1973—1974), S 121-136.

Roggenkamp, Martha M./Woerdman, Edwin: Legal Design of Carbon Capture and Storage—Developments in the Netherlands from an Internatioanal and EU Perspective, Intersentia, 2009.

Rothwell, Donald R./Stephens, Tim: The International Law of the Sea, Hart Publishing, 2010.

Ross, David A.: Opportunities and Uses of the Ocean, Springer, 1978.

Ross, David/Fenwick, Judith: Marine Scientific Research: US Perspective of

Jurisdiction and International Cooperation, in Lewis M. Alexander/Scott Allen/Lynne Carter Hanson (eds.), New Development in Marine Science and Technology: Economic, Legal and Political Aspects of Change, The Law of the Sea Institute, University of Hawaii, 1988, S 308-318.

Rothwell, Donald R. /Stephens, Tim: The International Law of the Sea, Oxford Hart Publishing, 2010.

Salamone, A. Xerri: The International Legal Regime of Installations at Sea, in Budislav Vukas (ed.), Essays on the New Law of the Sea 2, Zagreb, 1990, S431-449.

Sand, Peter H.: Principles of International Environmental Law, Cambridge, 2003.

Schachter, Oscar: International Law in Theory and Practice, Martinus Nijhoff Publishers, 1991.

Scholz, Wesley S.: Oceanic Research—International Law and National Legislation, Marine Policy, 4 (1980), S 91-125.

Scott, S. V.: The Inclusion of Sedentary Fisheries within the Continental Shelf Doctrine, International and Comparative Law Quarterly, 41 (1992), S 788-807.

Scovazzi, Tullio: Mining, Protection of the Environment, Scientific Research and Bioprospecting: Some Considerations on the Role of the International Sea-Bed Authority, IJMCL, 19 (2004), S 383-409.

Škrk, M.: The Prospects of Marine Scientific Research in the Contemporary Practice of States, in Budislav Vukas (ed.), Essays on the New Law of the Sea 2, ZAGREB, 1990, S 341-368.

Soons, A. H. A.: Artificial Islands and Installations in International Law, University of Rhode Island Occasional Paper, 22 (1974).

—. The Legal Regime of Marine Scientific Research: Current Issues, in Myron H. Nordquist/Ronan Long/Tomas H. Heidar/ John Norton Moore (eds.), Law, Science & Ocean Management, Martinus Nijhoff Publishers, 2008, S 139-166.

—. Marine Scientific Research and the Law of the Sea, The Hague, 1982.

—. Regulation of Marine Scientific Research by the Europe Community and Its Member States, ODIL, 23 (1992), S 259-277.

—. Freedom of Scientific Research, in Lewis M. Alexander/Scott Allen/Lynne Carter Hanson (eds.), New Development in Marine Science and Technology: Economic, Legal and Political Aspects of Change, The Law of the Sea Institute, University of Hawaii, 1988, S 293-307.

Side, J. /Baine, M. /Hayes, K.: Current Controls for Abandonment and Disposal of Offshore Installations at Sea, Marine Policy, 17 (1993), S 354-362.

Steinhoff, T. et al.: Estimating Mixed Layer Nitrate in the North Atlantic Ocean, Biogeosciences, 7(2010), S 795-807.

Stel, Jan H.: Towards a European Office for Marine Science and Technology, Marine Policy, 17 (1993), S 309-321.

Sherman, Kenneth: Large Marine Ecosystems: A Case Study, in Lewis M. Alexander/Scott Allen/Lynne Carter Hanson (eds.), New Development in Marine Science and Technology: Economic, Legal and Political Aspects of Change, The Law of the Sea Institute, University of Hawaii 1988, S 97-114.

Telszewski, M. et al.: Estimating the Monthly pCO_2 Distribution in the North Atlantic Using a Self-organizing Neural Network, Biogeosciences, 6 (2009), S 1405-1421.

Theutenberg, Bo Johnson: The Evolution of the Law of the Sea, Tycooly International Publishing Limited, 1984.

Tomczak, M.: Defining Marine Pollution, Marine Policy, 8 (1984), S 311-322.

Treves, Tullio: Principles and Objectives of the Legal Regime Governing Areas beyond National jurisdiction, Erik J. Molenaar and Alex G. Oude Elferink (eds.), The International Legal Regime of Areas Beyond National Jurisdiction: Current and Future Developments, Martinus Nijhoff Publishers, 2010, S 7-25.

—. The Seabed Beyond the Limits of National Jurisdiction: General and Institutional Aspects, Erik J. Molenaar and Alex G. Oude Elferink (eds.), The International Legal Regime of Areas Beyond National Jurisdiction: Current and Future Developments, Martinus Nijhoff Publishers, 2010, S 43-60.

Tsujino, Teruhis: Exploration Technologies for the Utilization of Ocean Floor Resources: Contribution to the Investigation for the Delineation of Continental Shelf, Science & Technology Trends Quarterly Review, 24 (2007), S 68-80.

Varghese, Peter: Law Enforcement Capabilities in EEZs: Australia and the South Pacific Island States, in Lewis M. Alexander/Scott Allen/Lynne Carter Hanson (eds.), New Development in Marine Science and Technology: Economic, Legal and Political Aspects of Change, The Law of the Sea Institute, University of Hawaii, 1988, S 150-152.

Verlaan, Philomene: Marine Scientific Research: Its Potential Contribution to Achieving Responsible High Seas Goverance, IJMCL, 27 (2012), S 805-812.

—. Current Legal Developments London Convention and London Protocol, IJMCL, 26 (2011), S 185-194.

—. Geo-engineering, the Law of the Sea, and the Climate Change, CCLR, 2009, S 446-458.

Vinuales, Jorge E.: Legal Techniques for Dealing with Scientific Uncertainty in Envrionmental Law, VJTL, 43 (2010), S 437-503.

Virgoe, John: International Governance of a Possible Geoengineering Intervention to Combat Climate Change, Climate Change, 95 (2009), S 103-119.

Vivian, Chris.: Marine Geoengineering—International Governance Issues, International Ocean Stewardship Forum 2009, aviliable at http://www.oceanstewardship.com/IOSF%202009/Keynotes_2009/CVivian_2009.pdf (last visited on 23 April 2013).

Vogel, Rainer: Flag States and New Registries, in Alastair D. Couper (ed.), The Marine Environment and Sustainable Development: Law, Policy and Science (1993), S 421.

Vukas, Budislav: The Law of the Sea—Selected Writings, Martinus Nijhoff Publishers, 2004.

—. Military Uses of the Sea and the United Nations Law of the Sea Convention, in Budislav Vukas (ed.), Essays on the New Law of the Sea 2, Zagreb 1990, S401-427.

Watson, Andrew J. et al.: Tracking the Variable North Atlantic Sink for Atmospheric CO_2, Science, 326 (2009), S 1391-1393.

Warner, Robin: Protecting the Diversity of the Depths: Environmental Regulation of Bioprospecting and Marine Scientific Research beyond National Jurisdiction, Ocean Yearbook, 22 (2008), 411-443.

Webster, Ferris: Technology and Data Management: US Science Community Views, in Lewis M. Alexander/Scott Allen/Lynne Carter Hanson (eds.), New Development in Marine Science and Technology: Economic, Legal and Political Aspects of Change, The Law of the Sea Institute, University of Hawaii, 1988, S 345-351.

Wegelein, Florian H. Th.: Marine Scientific Research, Martinus Nijhoff Publishers, 2005.

White, Angelicque et al.: An Open Ocean Trial of Controlled Upwelling Using Wave Pump Technology, Journal of Atmospheric and Oceanic Technology, 27 (2010), S 385-396.

Withee, Gregory W./Hamilton, Douglas R.: Opportunities in Oceanographic Science Offered by New Advances in Data Management, in Lewis M. Alexander/Scott Allen/Lynne Carter Hanson (eds.), New Development in Marine Science and Technology: Economic, Legal and Political Aspects of Change, The Law of the Sea Institute, University of Hawaii, 1988, S 322-332.

Wolfrum, Rüdiger/Matz, Nele: Conflicts in International Environmental Law, Berlin, 2013.

Wolfrum, Rüdiger: The Emerging Customary Law of Marine Zones: State Practice and the Convention on the Law of the Sea, NYIL XVIII, 1987, S121-144.

Wood, Michael: International Seabed Authority (ISA), in Rüdiger Wolfrum (ed.), Max Planck Encyclopedia of Public International Law, online edition.

Wooster, Warren S./Redfield, Michael: Consequences of Regulating Oceanic Research, in Warren S. Wooster (ed.), Freedom of Oceanic Research: A Study Conducted by the Center for Marine Affairs of the Scripps Institution of Oceanography, Crane, Russak&Company, Inc., 1973, S 219-234.

—. Scientific Aspects of Maritime Sovereign Claims, ODIL, 1 (1973), S 13-20.

Yang, Haijiang: Jurisdiction of the Coastal State over Foreign Merchant Ships in International Waters and the Territorial Sea, Springer, 2005.

Yusuf, Abdulaqwi A.: Toward a New Legal Framework for Marine Research: Coastal-State Consent and International Coordination, VJIL, 19 (1979), S 411-429.

图书在版编目(CIP)数据

现代科技背景下海洋法中的海洋科学研究＝Marine scientific research under the international law of the sea in the era of marine high-tech/常虹著. —厦门：厦门大学出版社，2014.9
ISBN 978-7-5615-5197-4

Ⅰ. ①现… Ⅱ. ①常… Ⅲ. ①海洋学-研究 Ⅳ. ①P7

中国版本图书馆 CIP 数据核字(2014)第 181115 号

官方合作网络销售商：

厦门大学出版社出版发行

(地址:厦门市软件园二期望海路 39 号　邮编:361008)
总编办电话:0592-2182177　传真:0592-2181253
营销中心电话:0592-2184458　传真:0592-2181365
网址:http://www.xmupress.com
邮箱:xmup @ xmupress.com

厦门大嘉美印刷有限公司印刷

2014 年 9 月第 1 版　2014 年 9 月第 1 次印刷
开本:720×1000　1/16　印张:14.5　插页:2
字数:260 千字
定价:46.00 元
本书如有印装质量问题请直接寄承印厂调换